A Mathematician
Reads the Newspaper

A Mathematician Reads the Newspaper

John Allen Paulos

ANCHOR BOOKS
DOUBLEDAY
New York London Toronto Sydney Auckland

AN ANCHOR BOOK
PUBLISHED BY DOUBLEDAY
a division of Bantam Doubleday Dell Publishing Group, Inc.
1540 Broadway, New York, New York 10036

ANCHOR BOOKS, DOUBLEDAY, and the portrayal of an anchor are trademarks
of Doubleday, a division of Bantam Doubleday Dell Publishing Group, Inc.

A Mathematician Reads the Newspaper was originally published
in hardcover by Basic Books in 1995. The Anchor Books edition
is reprinted by arrangement with Basic Books, a division of
HarperCollins Publishers, Inc.

Designed by Ellen Levine

Library of Congress Cataloging-in-Publication Data

Paulos, John Allen.
A mathematician reads the newspaper/John Allen Paulos.
—1st Anchor pbk. ed.
p.   cm.
Includes bibliographical references (p.205–6) and index.
1. Mathematics—Popular works.   I. Title.
QA93.P385   1996
510—dc20   95-46049
CIP

ISBN 0-385-48254-x
Copyright © 1995 by John Allen Paulos
All Rights Reserved
Printed in the United States of America
First Anchor Books Edition: April 1996

10   9

To storytelling number-crunchers and number-crunching storytellers

Contents

Section 2:
LOCAL, BUSINESS, AND SOCIAL ISSUES

Section 5:
FOOD, BOOK REVIEWS, SPORTS, OBITUARIES

Introduction

"I read the news today, Oh Boy."
—JOHN LENNON

My earliest memories, dating from the late 1940s, include hearing a distant train whistle from the back steps of the building we lived in on Chicago's near north side. I can also see myself crying under the *trapezi* (Greek for "table") when my grandmother left to go home to her apartment. I remember watching my mother rub her feet in bed at night, and I remember my father playing baseball and wearing his baseball cap indoors to cover his thinning hair. And, lest you wonder where I'm heading, I can recall watching my grandfather at the kitchen table reading the *Chicago Tribune*.

The train whistle and the newspaper symbolized the outside world, frighteningly yet appealingly different from the warm family ooze in which I was happily immersed. What was my grandfather reading about? Where was the train going? Were these somehow connected?

When I was five, we moved from a boisterous city block to the sterile environs of suburban Milwaukee, 90 miles and 4 light-years to the north. Better, I suppose, in some conventional 1950s sense, for my siblings and me, but it never felt as nurturing, comfortable, or alive. But this introduction is not intended to be an autobiography, so let me tell you about the *Milwaukee Journal*'s Green Sheet. This insert, literally green, was full of features that fascinated me. At the top was a saying by Phil Osopher that always contained some wonderfully puerile pun. There was also the "Ask Andy" column: science questions and brief answers. Phil and Andy became friends of mine. And then there was an advice column by a woman with the unlikely name of Ione Quinby Griggs, who gave no-nonsense Midwestern counsel. Of course, I also read the sports pages and occasionally even checked the first section to see what was happening in the larger world.

Every summer my siblings and I left Milwaukee and traveled to Denver, where my grandparents had retired. On long, timeless Saturday afternoons, I'd watch Dizzy Dean narrate the baseball game of the week on television and then listen through the static on my grandmother's old radio-as-big-as-a-refrigerator as my hero, Eddie Matthews, hit home runs for the distant Milwaukee Braves. The next morning I'd run out to the newspaper box on the corner of Kierney and Colfax, deposit my 5 cents, and eagerly scour the *Rocky Mountain News* for the box scores. A few years later, I would scour the same paper for news of JFK.

Back home, my affair with the solid *Milwaukee Journal* deepened (local news, business pages, favorite columnists) until I left for the University of Wisconsin in Madison, at which time the feisty *Capitol Times* began to alienate my affections. Gradually my attitude toward newspapers matured and, upon moving to Philadelphia after marriage and graduate school, my devotion devolved into a simple adult appreciation of good newspaper reporting and writing. My former fetishism is still apparent, however, in the number of papers I read and in an excessive affection for their look, feel, smell, and peculiarities. I subscribe to the *Philadelphia Inquirer* and to the paper of record, the *New York Times*, which arrives in my driveway wrapped in blue plastic. I also regularly skim the *Wall Street Journal* and the *Philadelphia Daily News*, occasionally look at *USA Today* (when I feel a powerful urge to see weather maps in color), the *Washington Post*, the suburban *Ambler Gazette*, the *Bar Harbor Times*, the local paper of any city I happen to be visiting, the tabloids, and innumerable magazines.

At fairly regular intervals and despite the odd credential of a Ph.D. in mathematics, I even cross the line myself to review a book, write an article, or fulminate in an op-ed. But if I concentrate on it, reading the paper can still evoke the romance of distant and uncharted places.

One result of my unnatural attachment to newspapers is this book. Structured like the morning paper, *A Mathematician Reads the Newspaper* examines the mathematical angles of stories in the news. I consider newspapers not merely out of fondness, however. Despite talk of the ascendancy of multimedia and the decline of print media, I think the rational tendencies that newspapers foster will survive (if we do), and that in some form or other newspapers will remain our primary means of considered public discourse. As such, they should enhance our role as citizens and not reduce it to that of mere consumers and voyeurs (although there's nothing wrong with a little buying and peeking). In addition to placing increased emphasis on analysis, background, and features, there is another, relatively unappreciated way in which newspapers can better fulfill this responsibility. It is by knowledgeably reflecting the increasing mathematical complexity of our society in its many quantitative, probabilistic, and dynamic facets.

This book provides suggestions on how this can be done. More important, it offers novel perspectives, questions, and recommendations to coffee drinkers, straphangers, policy makers, gossip mongers, bargain hunters, trendsetters, and others who can't get along without their daily paper. Mathematical naïveté can put such readers at a disadvantage in thinking about many issues in the news that may seem not to involve mathematics at all. Happily, a sounder understanding of these issues can be obtained by reflecting on a few basic mathematical ideas, and even those who despised the subject in school will, I hope, find them fascinating, rewarding, and accessible here.

But perhaps you need a bit more persuasion. Pulitzer, after all, barely fits in the same sentence as Pythagoras. Newspapers are daily periodicals dealing with the changing details of everyday life, whereas mathematics is a timeless discipline concerned with abstract truth. Newspapers deal with mess and contingency and crime, mathematics with symmetry and necessity and the sublime. The newspaper reader is everyman, the mathematician an elitist. Furthermore, because of the mind-numbing way in which mathematics is generally taught, many people have serious misconceptions about the subject and fail to appreciate its wide applicability.

It's time to let the secret out: Mathematics is not primarily a matter of plugging numbers into formulas and performing rote computations. It is a way of thinking and questioning that may be unfamiliar to many of us, but is available to almost all of us.

As we'll see, "number stories" complement, deepen, and regularly undermine "people stories." Probability considerations can enhance articles on crime, health risks, or racial and ethnic bias. Logic and self-reference may help to clarify the hazards of celebrity, media spin control, and reportorial involvement in the news. Business finance, the multiplication principle, and simple arithmetic point up consumer fallacies, electoral tricks, and sports myths. Chaos and nonlinear dynamics suggest how difficult and frequently worthless economic and environmental predictions are. And mathematically pertinent notions from philosophy and psychology provide perspective on a variety of public issues. All these ideas give us a revealing, albeit oblique, slant on the traditional Who, What, Where, When, Why, and How of the journalist's craft.

The misunderstandings between mathematicians and others run in both directions. Out of professional myopia, the former sometimes fail to grasp the crucial element of a situation, as did the three statisticians who took up duck hunting. The first fired and his shot sailed six inches over the duck. Then the second fired and his shot flew six inches below the duck. At this, the third statistician excitedly exclaimed, "We got it!"

Be warned that, although the intent of this book is serious and its tone largely earnest, a few of the discussions may strike the reader as similarly off the mark. Nonetheless, the duck hunters (and I) will almost always have a point. My emphasis throughout will be on qualitative understanding, pertinence to daily life, and unconventional viewpoints. What new insights does mathematics give us into news stories and popular culture? How does it obscure and intimidate? What mathematical/psychological rules of thumb can guide us in reading the newspaper? Which numbers, relations, and associations are to be trusted, which dismissed as coincidental or nonsensical, which further analyzed, supplemented, or alternatively interpreted? (Don't worry about the mathematics itself. It is either elementary or else is explained briefly in self-contained portions as needed. If you can find the continuation of a story on page B16, column 6, you'll be okay.)

The format of the book will be loosely modeled after that of a standard newspaper, not that of a more mathematical tome. I'll proceed through such a generic paper (*The Daily E^{X}ponent* might be an appropriate name) in a more or less linear manner, using it as a convenient lens through which to view mathematically various social concerns and phenomena. Not the least of these are newspapers themselves. The book will begin with section 1 news, including national and international stories, serious articles on politics, war, and economics, and the associated punditry. Then I move on to a variety of local, business, and social issues, then to the self, lifestyle, and soft-news section. After a discussion of reporting on science, medicine, and the environment, I'll conclude with a brief look at various newspaper features such as obituaries, book reviews, sports, advice columnists, top-ten lists, and so on.

Each section of *A Mathematician Reads the Newspaper* is composed of many segments, all beginning with a headline. So as not to fall victim to Janet Cooke's sin, I hereby acknowledge that these headlines are composites, invented to stimulate recall of a number of related headlines (most from 1993 or 1994, but of perennial concern). The segment will consider some of the pertinent underlying mathematics and examine how it helps explicate the story. Occasionally the tone will be debunking, as when I discuss the impossible precision of newspaper recipes that, after vague directions and approximate ingredient amounts, conclude happily that each serving contains 761 calories, 428 milligrams of sodium, and 22.6 grams of fat.

The mathematics will frequently suggest an alternative viewpoint or clarification. Incidence matrices, for example, provide society-page readers with a new tool for conceiving of the connections among the attendees at the Garden Club gala. And complexity theory helps elucidate the idea of the compressibility of a news story and the related notion of one's complexity horizon; some things, it happens, are too complicated for any of us to grasp. On a more prosaic level, claims were recently made that blacks in New York City vote along racial lines more than whites do. The evidence cited was that 95 percent of blacks voted for (black) mayor David Dinkins, whereas only 75 percent of whites voted for (white) candidate (and victor) Rudolph Giuliani. This assertion failed to take into account, however, the preference of most black voters for any Democratic candidate. Assuming that 80 percent of blacks usually vote for Democrats and that only 50 percent of whites usually vote for Republicans, one can argue that only 15 percent of blacks voted for Democrat Dinkins based on race and that 25 percent of whites voted for Republican Giuliani based on race. There are, as usual at the politico-mathematical frontier, countless other interpretations.

In showing up the interplay between mathematics and popular culture, I will digress, amplify, wax curmudgeonly, and muse regularly enough to establish a conversationlike environment, but I'll try not to be too cloying or pontifical in the process. The mathematical expositions, illustrations, and examples will be embedded in a sequence of largely independent news segments and thus, I trust, will not be threatening or off-putting. My aim is to leave the reader with a greater appreciation of the role of mathematics in understanding social issues and with a keener skepticism of its uses, nonuses, misuses, and abuses in the daily paper.

Despite his limited conception of mathematics, Samuel Johnson would have understood the point. Boswell quotes him as saying, "A thousand stories which the ignorant tell, and believe, die away at once when the computist takes them in his gripe."

Section 1

POLITICS, ECONOMICS, AND THE NATION

You can only predict things after they've happened.
—EUGENE IONESCO

I find it oppressive when a piece of writing has a single thesis that is stated early and is then continuously and predictably amplified and repeated. It reminds me of being cornered at a party by someone with interminably boring stories to tell who refuses to omit any detail or to deviate one jot from his sequential presentation (and I'm not using the pronoun "his" generically here). By contrast, part of a newspaper's appeal for me is its jumbled heterogeneity and random access. If I want to check the book reviews or celebrity features or health news or crime reports before I read about the Fed's raising the discount rate, then I do. I paid for the paper. Similarly, I've designed this book to allow those of you who first glance at other sections of the paper before returning to the front page to do the same here.

The news topics I treat in this first section include the economy (especially the laughable assumption that its nonlinear complexity is subject to precise prediction), war, conspiracy theories, high-stakes bluffing, and political power and abuses. Also discussed are ambiguous language, the inverted pyramid structure of news stories, a few relevant psychological findings, and, of course, a bit of mathematics.

I begin with some issues involved in the making of social choices. How do we weigh alternatives? How do we settle issues by ballot? How do we distribute goods? The necessity for such choices follows from, among other things, the fact that our two most basic political ideals—liberty and equality—are, in their purest forms, incompatible. Complete liberty results in inequality, and mandatory equal-

ity leads to a loss of liberty. Today's *New York Times* headline, HOUS-
ING RIGHTS VIE WITH FREE SPEECH, aptly attests to this. How to
apportion common assets among contending parties is another classic problem that is amply illustrated in the newspaper. PUBLIC SQUAB-
BLE OVER HARRIMAN ESTATE is one recent instance.

A glimmer of the mathematical facets of such matters can be
detected in the joke about two brothers arguing over a large piece of
chocolate cake. The older brother wants it all; the younger one wails
that this isn't fair–the cake should be split 50-50. The mother enters
and makes them compromise. She gives three-fourths of the cake to
the older brother and one-fourth to the younger. The story takes on
a somber resonance if we identify the older brother with Serbia, the
younger with Bosnia, and the mother with the Western powers.

There is, of course, a better apportionment scheme for dividing a
cake fairly: one brother cuts the cake, and the other chooses which
part to take. No need for mother. This isn't going to happen in
Bosnia or New York, but as a cerebral warm-up for the first segment,
you might want to ponder how to generalize this procedure. Imagine
that Mom bakes a large cake and calls in her hungry brood. How
should her four children, George, Martha, Waldo, and Myrtle, go
about dividing the cake evenly among themselves without her intervention?*

---

*George slices off what he considers to be a quarter of the cake. If Martha judges the piece
to be a quarter of the cake or less, she doesn't touch it. If she thinks it's bigger than a quarter of the cake, she shaves off a sliver to make it exactly a quarter. Waldo then either leaves
the piece alone or trims it further if he thinks it's still bigger than a quarter of the cake.
Finally, Myrtle has the same option: trim it if it's too big and leave it alone if it's not. The
last person to touch the slice keeps it. (But what is to guard against each person cutting too
small or too large a piece?) This finished, there are three people remaining who must divide
the remainder of the cake evenly. The same procedure is followed. The first person slices
off what he or she considers to be a third of the remaining cake (equivalent to a quarter of
the original) and so on. In this way everyone is convinced that he or she has received a
quarter of the cake.

Lani "Quota Queen" Guinier

Voting, Power, and Mathematics

Vilified as a "quota queen" and hailed as an activist superwoman, Lani Guinier probably became a greater news presence than she would have if President Clinton's nomination of her as assistant attorney general for civil rights had been approved by the Senate. Most of us would be hard-pressed to come up with the name of the present occupant of that position. I'm sympathetic to (most of) the aims of the Voting Rights Act, yet strongly opposed to quotas (whether they're called that or not)—but rather than rehash the ideological aftermath of the political fray, let me describe a simple mathematical idea that motivates some of Professor Guinier's writings. It is the Banzhaf power index, named after a lawyer, John F. Banzhaf, who introduced it in 1965.

Imagine a small company with three stockholders. Assume that these stockholders hold, respectively, 47 percent, 44 percent, and 9 percent of the stock, a simple majority of 51 percent being necessary to pass any measure. It's clear, I think, that although one of them may drive a Yugo, all three stockholders have equal power. That's because any two of them are sufficient to pass a measure.

Now consider a corporation with four stockholders holding, respectively, 27 percent, 26 percent, 25 percent, and 22 percent of the stock. Again a simple majority is needed to pass any measure. In this case any two of the first three stockholders can pass a measure, whereas the last stockholder's vote is never crucial to any outcome. (When the last stockholder's 22 percent is added to any one of the first three stockholders' percentages, the sum is less than 51 percent, and any larger coalition of stockholders doesn't require the last stockholder's 22 percent.) The last stockholder is called a *dummy*, an apt technical term for someone whose vote can never change a losing coalition into a winning one or vice versa. The dummy has no power; the other three stockholders have equal power. (Incidentally, the *Wall Street Journal*, which led the attack on Ms. Guinier, should have appreciated the prevalence of unconventional voting schemes in business.)

One more example before the definition. Imagine that representatives to the national assembly of the new country of Perplexistan split along ethnic lines—45 percent, 44 percent, 7 percent, and 4 percent, respectively. Any two of the first three ethnic groups may form a majority coalition, but the smallest party is a dummy. Thus, despite the fact that the third group's representation is much smaller than that of the first two groups and only slightly larger than that of the fourth group, the first three groups have equal power, and the last has none.

The Banzhaf power index of a group, party, or person is defined to be the number of ways in which that group, party, or person can change a losing coalition into a winning coalition or vice versa. I've examined only cases where the parties with any power at all share equal power, but with the definition in hand more complicated cases may be discussed.*

There have been a number of schemes suggested to ensure that a group's power as measured by the Banzhaf index more closely reflects its percentage of the vote. This may be a special concern when a minority's interests are distinct from those of a biased major-

---

*Consider a company or political body in which four parties—let's be romantic and call them A, B, C, and D—have 40 percent, 35 percent, 15 percent, and 10 percent of the vote, respectively. If one methodically lists all possible voting situations (A, C, and D for, B against; B and D for, A and C against; and so on), one finds that there are ten of them in which party A's vote is pivotal (changes a winning coalition into a losing one or vice versa), six in which B's vote is, six in which C's is, and only two in which D's vote is pivotal. Thus the parties' respective power indices are 10, 6, 6, 2, indicating that party A is five times as powerful as party D, while parties B and C have equal power and are only three times as powerful as party D. There are no dummies.

ity that retains all the power in a given district. When this happens to be the case in some district, a somewhat different proposal put forward by Ms. Guinier would grant to each voter a number of votes equal to the number of contested seats in the district. Under this so-called cumulative voting procedure, the voter could distribute his or her votes among the candidates, spreading them about or casting them all for a single representative. Although animated by a desire to strengthen the Voting Rights Act and facilitate the election of minority representatives, this proposal need make no essential reference to race and would help any marginal group to organize, form coalitions, and attain some power.

Imagine a city council election in which five seats are at stake and there are a large number of candidates for them. Instead of the standard procedure of dividing the city into districts and having each district elect its representative to the council, cumulative voting would grant every voter five votes to distribute among the candidates as he or she wished. If *any* group of voters was committed and cohesive enough, they could cast all five of their votes for a single candidate whose interests would reflect theirs. Just such a proposal has been broached as a substitute for congressional districts that have been racially gerrymandered to allow for the election of African-American congressmen. An article in the *New York Times* in April 1994 suggested a way to replace this unappealing balkanization. North Carolina, home of the snakelike 12th Congressional District, might seriously consider dividing the state along natural geographic lines: the Eastern, Piedmont, and Western regions. Within each of these regions, which are presently home to four, five, and three representatives, respectively, cumulative voting would be instituted.

Such tinkering with election procedures is not unheard of. In various counties in New York State, for example, there are voting systems in which representatives' votes are weighted to make power accord with population and to ensure that no representative is a dummy in the technical sense. (The standard sort is harder to eliminate.) The recent effort to impose congressional term limits is another instance, as are various sorts of sequential runoffs, requirements for super majorities, and so-called Borda counts, whereby voters rank the candidates and award progressively more points to those higher in their rankings. (Proponents of change sometimes tendentiously frame the issue by saying that 51 percent of the vote results in 100 percent of the power. Opponents never bring up the parliamentary systems in Europe and Israel, which frequently allow 1 percent of the vote to establish a critical seat in Parliament.)

Approval voting is yet another system that might be appropriate in certain situations, primary elections in particular. In this case each voter chooses, or approves of, as many candidates as he or she wants. The principle of "one person, one vote" is replaced with "one candidate, one vote," and the candidate receiving the greatest number of approvals is declared the winner. Scenarios in which, for example, two liberal candidates split the liberal vote and allow a conservative candidate to win with 40 percent of the electorate would not arise. (Can you think of any drawbacks to approval voting, however?)

The U.S. Senate, where the disproportionate clout of less populous states constitutes a significant, if almost invisible, deviation from pure majority rule, is not immune to such anomalies. The fact is that every voting method has undesirable consequences and fault lines (this is even a formal mathematical theorem, thanks to the economist Kenneth J. Arrow). Not whether but *how* we should be democratic is the difficult question, and an open experimental approach to it is entirely consistent with an unwavering commitment to democracy. Politicians who are the beneficiaries of a particular electoral system naturally wrap themselves in the mantle of democracy. So do would-be reformers. Lani Guinier's writings, the mathematical roots of which go back to the eighteenth century, remind us that this mantle can come in many different styles, all of them with patches.

Let me close with a tangential question suggested by news stories accompanying recent appointments to the Supreme Court. These stories often speculate about the possibility of a centrist bloc that could dictate decisions before the court. In fact, although each Supreme Court justice has equal power, a *cohesive* group of five justices could determine every ruling and, in effect, disenfranchise the other four and render them dummies. All that would be necessary would be for the five first to vote surreptitiously among themselves, determine what a majority of them thinks, and then agree to be bound by their secret ballot and vote as a bloc in the larger group. Can you think of some scenario whereby three of these five justices could determine court decisions?*

---

*If three members of the cabal (a subcabal, if you will) meet secretly beforehand, determine what a majority of them thinks, and then agree to be bound by this ballot and vote as a bloc in the larger cabal, they can determine the larger group's decision, which will, in turn, determine the decision of the whole court.

Bosnia: Is It Vietnam or World War II?

Psychological Availability and Anchoring Effects

The psychological literature contains many papers on the so-called availability error, a phenomenon I believe to be particularly widespread in the media. First described by the psychologists Amos Tversky and Daniel Kahneman, it is nothing more than a strong disposition to make judgments or evaluations in light of the first thing that comes to mind (or is "available" to the mind).

Are there more words having "r" as a first letter or as a third letter? What about "k"? Most people incorrectly surmise that more words have these letters in the first position than the third, since words such as *rich, real,* and *rambunctious* are easier to recall (*recall* is another) in this context than are words such as *fare, street, throw,* and *words.*

Another case derives from a group of people asked by psychologists to memorize a collection of words that included four terms of praise: *adventurous, self-confident, independent,* and *persistent.* A second group is asked to memorize a similar list, except that those four positive words are replaced by *reckless, conceited, aloof,* and *stubborn.* Both groups then move on to an ostensibly different task—reading a somewhat ambiguous news story about a young man whom they are then asked to evaluate. The first group thinks much more highly of the young man than does the second, presumably because the positive words they have just memorized are more available to them. (Anyone can exploit these temporary associations. After noting that olive oil comes from squeezing olives, palm oil from compressing palm fruit, and peanut oil from mashing peanuts, Lily Tomlin has inquired about the source of baby oil.)

And when read a list of names, half male, half female, people will judge most of the names to have been male if some of the male names are famous ones. On the other hand, if some of the female names are famous (and none of the male ones were), they will judge that most of the names read are female.

As I will discuss in the next section, material that has been recently presented or is emotionally evocative, dramatic, or concrete is generally more available than that which is old, emotionally neutral, boring, or abstract. The vividness of a car hijacking or a product tampering makes such events much more worrisome than they should be. In the other direction, deaths from stroke and heart disease are so commonplace that they almost go unnoticed. Casinos are another good example. Contrast the bells and lights that accompany even the smallest wins with the soundless invisibility of even the largest losses.

Or consider the wording of poll questions. The Yankelovich polling organization asked the following question (not devised by them, but taken from a Ross Perot questionnaire): Should laws be passed to eliminate all possibilities of special interests giving huge sums of money to candidates? To this, 80 percent of the sample responded yes, 17 percent no. The same organization then posed the question: Should laws be passed to prohibit interest groups from contributing to campaigns, or do groups have a right to contribute to the candidate they support? In this case only 40 percent said Yes, while 55 percent answered no. Presumably, the latter, less provocative question made the opposing argument more available to the respondents.

Pick up any newspaper and there are bound to be national stories whose reception is strongly colored by past stories to which they bear only a superficial resemblance, but which are psychologically available. A newspaper includes an account of the situation in Bosnia that almost explicitly calls on its readers to note the similarities to the Holocaust. To counter, a government official brings up another tenuously related epoch, the quagmire of Vietnam (the last three words are used so often that they should be hyphenated). Because President and Mrs. Clinton's association with Whitewater loosely fits the category "presidential scandal," the parallels with Watergate are brought to mind. A story informs us that a majority of Americans are in favor of the caning of an American teenager convicted of vandalism in Singapore. The cruelty of the practice is seen against, and somewhat masked by, the runaway crime rates in this country. It's tempting to dismiss the caning by contrasting it with the murders of foreign tourists here.

When and in what company a story breaks also greatly influences our interpretation of it, and this, I stress, is largely a matter of luck. Has it been a slow news day? Or has some riveting narrative muted everything else (as the O. J. Simpson megastory did with Jimmy Carter's peace mission to North Korea)? Does the story resonate with another recent story? Or does it recall an old story that the public reacted to in a uniformly positive or negative way? The clubbing of the Olympic skater Nancy Kerrigan brought to mind the stabbing of the tennis star Monica Seles and the stalking of some well-known actresses, and thus we were predisposed to believe that the clubber was a sports fanatic obsessed with Ms. Kerrigan.

The availability error glides imperceptibly into deeper waters. Models that are fitting for one domain are frequently inappropriate for another that is superficially similar. Competition between companies, say, Coke and Pepsi, is taken as the model for competition between countries, say, the United States and Japan. But, as the economist Paul Krugman has pointed out, this analogy doesn't hold water (or cola). Only a small fraction of the sales of Pepsi and Coke goes to their own workers. One's gain is often the other's loss. The American economy, even in these interconnected times, produces goods and services that are largely (90 percent) for its own use, and it is not necessarily hurt by the success of the Japanese economy. International trade is far from a zero-sum game.

There is no foolproof way for readers to avoid the tendency to be taken in by facile parallels and analogies. More complex coverage would help, but there are necessary limitations to this approach. A weaker but more practical antidote is consciously to search for interpretations or associations that undermine the prevailing one that is so temptingly available. Similar though they may be on the surface, how are the events in question actually different? How can we put the given story into a more neutral context? What's crucial and what's accidental about it? One might argue that it's pointless to struggle against our predisposition to search for resemblances; it's human nature, after all. The problem with such an argument is that it proves too much. Racism might also be considered natural, for example.

Finally, there are a couple of corollaries to the availability error that also have implications for journalists and readers. One is the *halo effect*: the tendency to judge a person or a group in terms of one salient characteristic. Research articles that are identical in all respects except for the prestigious university affiliation of the author have been submitted to scholarly journals with the predictable consequences: they're rejected. A similar fate has befallen novels by famous authors submitted to publishers under a pseudonym. The relevance of the halo effect to news reporting is obvious. It also partially explains why so many experts on *Nightline*, for example, have Harvard and Yale pedigrees.

Another related chink in our rationality goes by the name of *anchoring effects* and can be demonstrated by anyone willing to ask a few questions. It is illustrated by studies in which people were asked to estimate the population of Turkey. Before answering, they were presented with a figure and asked whether the actual number was higher or lower. Of those who were first presented with the figure of 5 million, the average estimate was 17 million; of those first presented with a figure of 65 million, the average estimate was 35 million. As Tversky, Kahneman, and others have established, people are "anchored" to the original figure presented to them and, although they move in the right direction away from it, they are reluctant to move too far. (The population of Turkey was approximately 50 million when the experiment was performed.) You might want to try the experiment yourself. Using the anchors of 2 million and 100 million, I had good results with Argentina, whose population is approximately 33 million.

In a related, more mathematical demonstration of anchoring effects, one set of people is asked to estimate $1 \times 2 \times 3 \times 4 \times 5 \times 6 \times 7 \times 8$, and another set is asked to estimate $8 \times 7 \times 6 \times 5 \times 4 \times 3 \times 2 \times 1$. The median value of the first group's guesses was 512, that for the second group 2,250. Again the anchoring effect, the insufficient adjustment upward or downward from an original value, explains the discrepancy. The first group starts multiplying 1, 2, 3, and so on, and the estimate is somewhat anchored to these lower numbers, whereas the second group begins with the larger numbers. (By the way, both groups seriously underestimated the product, which is 40,320.)

The bottom line (or paragraph): Uncritical news-gathering routines tend to bolster the conventional wisdom. They're too often anchored to "what everybody knows," to simplistic analogies, to whatever is psychologically available.

Recession Forecast If Steps Not Taken

Unpredictability, Chaos, and Pooh-Poohing the Pooh-Bahs

I find reading old newspaper analyses, government press releases, and bygone bits of punditry both sobering and entertaining. They often seem to presuppose that political and economic matters are, with a little thought and perhaps some calculation, more or less predictable. Obviously, such matters are not very predictable, and there are some surprising mathematical reasons why they are not.

These mathematical reasons ensure that much economic and political commentary and forecast are fatuous nonsense, no more on target than the farmer marksman with hundreds of chalked bull's-eyes on the wall of his barn, each with a bullet hole in its center. When asked how he could be so accurate, the farmer, who had perhaps read Ionesco, admitted that he first made the shot and then drew the bull's-eye around it.

In fact, stripped to their essence, many social forecasts may be paraphrased in one of two ways. The first is: "Things will continue roughly as they have been." When pressed, the pooh-bahs and prognosticators admit a further clause: "until something changes." The other way is equivalent, but puts the emphasis on the change: "Things will change." Here again, when pressed, the pooh-bahs and prognosticators admit a further clause, "after an indeterminate period of stability." But THINGS TO STAY SAME UNTIL CHANGE or THINGS TO CHANGE EVENTUALLY are too obviously hollow and unfalsifiable to be suitable for the headline of a news analysis or a columnist's essay. Their emptiness has to be disguised.

Consider the typical analysis of the economy. It generally isolates one or two factors (or their absence) as the cause of this or that malady. There is generally more sophistication in football play-by-play broadcasting.

The economic cornerstone of Reaganomics, the economist Arthur Laffer's appropriately named Laffer curve, provides a good example. Laffer and others made the obvious point that a 100 percent tax rate would bring in almost no revenue for the government; few of us would have any incentive to work if all our money were being confiscated by the government. At the other extreme, a 0 percent tax rate would clearly bring in no revenue for the government either. Furthermore, if the tax rate were very low, say, 3 percent, doubling it to 6 percent would almost double the government's revenues. If the rate were a little higher, however, around 15 percent, say, doubling it would not have such a pronounced effect; revenues would increase more modestly (see diagram).

Revenue versus Tax Rate

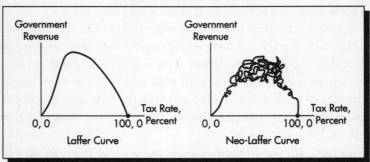

likewise, at the other end of the tax rate spectrum. A government whose tax rate was 97 percent would be larcenous, but if it lowered the rate to 94 percent, it would likely provide enough incentive for workers to increase their output and thus swell government revenues significantly. However, if the tax rate were lower, say, 85 percent, and the government decreased it to 70 percent, the increase in government revenue would likely not be as dramatic.

Revenues increase when high rates are lowered as well as when low rates are raised but, in each case, to a lesser degree as diminishing returns set in.

Now it only seems reasonable and a matter of geometric necessity that a tax rate somewhere between 0 and 100 percent should bring in the maximum government revenue. The result of this line of argument is the Laffer curve pictured in the illustration. Many people find the argument convincing and, believing that the economy is on the right side of the hump, argue that a decrease in tax rates would notably raise government revenues.

But is what happens in the middle of the curve really as clear as what happens at its extremes? Martin Gardner in a somewhat derisive *Scientific American* article constructed a neo-Laffer curve whose interior is a whirlpool of snarls and convolutions but whose extremes are nevertheless identical to those of the Laffer curve. Gardner's curve has many different points of maximum revenue and which one of them, if any, is achieved depends on an indeterminate number of historical and economic contingencies. These factors and their interaction are too complicated to be determined by variation in any single variable such as the tax rate.

In general, too little notice is taken of the interconnectedness of the variables in question. Interest rates have an impact on unemployment rates, which in turn influence revenues; budget deficits affect trade deficits, which sway interest rates and exchange rates; consumer confidence may rouse the stock market, which alters other indices; natural business cycles of various periods are superimposed on one another; an increase in some quantity or index positively (or negatively) feeds back on another, reinforcing (weakening) it and being in turn reinforced (weakened) by it. These and a myriad of more complicated interactions characterize the economy.

The notion of a nonlinear dynamical system can be used to model such interconnections and, more important for my purposes, helps clarify why we should not expect to predict political or economic developments with any exactitude. Before I define such systems, I want you to picture a more tangible artifact: a pool table. Imagine that approximately twenty to thirty round obstacles are securely fastened to it in haphazard placement (see diagram). Your challenge is to hire the best pool player you can find and ask him to place the ball at a particular spot on the table and take a shot toward one of the round obstacles. After he's done so, ask him to make exactly the same shot from the same spot with another ball. Even if his angle on this second shot is off by the merest fraction of a degree, the trajectories of these two balls will very soon diverge considerably. An infinitesimal difference in the angle of impact will be magnified by successive hits of the obstacles. Soon one of the trajectories will hit an obstacle that the other missed entirely, at which point all similarity between the two trajectories ends.

Billiards—Magnification of Small Differences

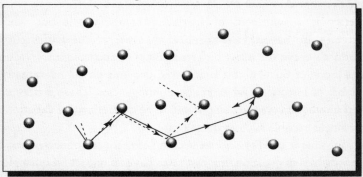

The sensitivity of the billiard balls' paths to minuscule variations in their initial angles is not totally unlike, say, the dependence of one's genetics on which zigzagging sperm cell reaches the egg. This chanciness also brings to mind the disproportionate effect of seemingly inconsequential events: the missed planes, serendipitous meetings, and odd mistakes that shape and reshape our lives. (A job recruiter I know once rejected a well-qualified French applicant because in a discussion of tabloids and pornography, he laughingly mentioned that his mother's picture had once appeared in *Peep Hole* magazine. She thought it was either true or a poor joke. Later she realized he meant *People* magazine.) The inevitable amplification of tiny differences in detail is just one of the factors suggesting that the economy is almost invulnerable to dependable forecast.

Technically, nonlinear dynamical systems are not billiard tables or economic systems, but mathematical spaces on which vector fields are defined. A vector field may be thought of as a *rule f* saying, in effect, that "if an object is currently at a point x, it moves next to point f(x), then to point f(f(x)), and so on." The rule f is nonlinear if, for example, the variables involved are squared or multiplied together and the sequence of the object's positions is its trajectory. A mathematical subterfuge allows us to consider the movement of a fictitious object around a very high dimensional space instead of the movement of many objects around a lower-dimensional space.

For rough expository purposes, we can collapse the important distinction between a mathematical model and a part of reality and think of a system as any collection of parts whose movements and interaction can be described by rules and/or equations, however imprecise. The postal service, the human circulatory system, the local ecology, and the operating system of the computer on which I write are all examples of this loose notion of a system. A nonlinear system is one whose elements—again, I'm writing quite loosely—are not linked in a linear or proportional manner. They are not linked, for example, as they are in a bathroom scale or a thermometer; doubling the magnitude of one part will not double that of another, nor is the output proportional to the input. Linear systems involve equations like $Z = 7X + 2Y$; nonlinear systems have equations like $Z = 5X^2 + 3XY$.

Chaos theory (and, to a lesser extent, the study of nonlinear systems) may be said to have been born in 1960 when the meteorologist Edward Lorenz was playing with a computer model of a simple weather system. Plugging numbers into his model, he derived a set of weather projections. Later, when he ran his program again, he plugged in numbers that had been rounded off to three decimal places instead of six and noticed that the resulting weather projections soon diverged from the original ones and that the two sets of projections soon bore no discernible relation to each other.

Although Lorenz's nonlinear model was simplistic (it involved only three equations and three variables) and his computer equipment was primitive, he drew the correct inference from this divergence of computer-simulated weathers: It was not a fluke. It was caused by the tiny variations in the system's initial conditions. In fact, the weather, even this simplified model of it, is not susceptible to precise long-range prediction because it, like the billiard table, is sensitive to almost imperceptible changes in the initial conditions. These changes lead to slightly bigger ones a minute later or a foot away, which lead to yet more substantial deviations, the whole process cascading over time into an aperiodic, nonrepetitive unpredictability. Of course, there are certain general constraints that are satisfied (no ice storms in Tanzania, little rain in the desert, rough seasonal temperature gradients, for example), but specific long-range forecasts are virtually worthless.

The sensitive dependence of nonlinear systems on their initial conditions has been called the Butterfly Effect, from the idea that a butterfly flapping its wings in China, say, might spell the difference several months later between a hurricane and a balmy day along the eastern U.S. seaboard.

Since Lorenz's work, there have been many manifestations of the Butterfly Effect in disciplines ranging from hydrodynamics (turbulence and fluid flow) to physics (nonlinear oscillators), from biology (heart fibrillations and epilepsy) to ecology (population changes). Furthermore, these nonlinear systems demonstrate a complex unpredictability that seems to arise even when the systems are defined by quite elementary, nonlinear rules and equations. Their trajectories in mathematical space are fractals, astoundingly intricate and self-similar.

A *fractal* is a wispy, tangled curve (or surface, solid, or higher-dimensional object) that contains more, but similar, complexity the closer one looks. The shoreline, to cite a standard example, has a typical jagged shape at whatever scale we draw it–that is, whether we use satellite photos to sketch the whole coast or the detailed observations of someone walking along a stretch of beach. Similarly, the surface of a mountain looks roughly the same whether seen from a height of 200 feet by a giant or up close by an insect. As the father of fractals Benoit Mandelbrot has stressed, clouds are not circular or elliptical, tree bark is not smooth, lightning does not travel in a straight line, and snowflakes are not hexagons. Rather, these and many other shapes in nature—the surfaces of battery electrodes, the spongy interior of intestines and lung tissue, the diffusion of a liquid through semiporous clays, the variation of commodities prices over time—are near fractals and have the characteristic zigzags, push/pulls, and bump/dents at almost every size scale, greater magnification yielding similar, but ever more complicated, convolutions.

This is all very interesting, but what lessons should newspaper readers take away from the discussion? I hope that even a loose, intu-itive understanding of the behavior of many interacting variables, the Butterfly Effect, complex nonlinear systems, positive and negative feedback, and so on, should arouse a certain wariness of glib and sim-plistic diagnoses. Our standard economic statistics are notoriously imprecise and unreliable, and this imprecision and unreliability prop-agates. The reader should be more mindful of stories where the effect of small differences—say, decreases in a couple of indices—seems to be magnified; of articles about the frenzied selling of stocks feeding self-fulfillingly upon itself; of accounts pointing to single causes or immediate consequences.

You should observe that the accuracy of social forecasts and predictions is vastly greater if the predictions are short-term rather than long-term; if they deal with simple rather than complex phe-nomena, with pairs of closely associated variables rather than many subtly interacting ones; if they're hazy anticipations rather than pre-cise assertions; and if they are not colored by the participants' inten-tions. Note how few political and economic predictions meet the conditions of these "ifs"—those are the ones to take seriously.

Lest by these brief suggestions I be accused of ignoring my own mis-givings about simplistic brevity (or of offering the paradoxical slogan: No More Slogans), I should mention that there is a body of research, both mathematical and empirical, that supports this counsel. Mathe-matical models of the economy, for example, are only awkwardly cast in a linear framework. The variables in realistic models interact in strongly nonlinear ways that give rise to the phenomena described here. Linear models are used regularly not because they are more accurate but because they are easier to handle mathematically. (Economists and physicists sometimes adopt the same research policy as the proverbial drunkard. Asked why he was looking for his keys under the streetlamp when he lost them up the block, the drunkard replied that the light was better there.)

Furthermore, empirical microeconomic studies by Mosekilde, Larsen, Sterman, Brock, LeBaron, Woodford, and others suggest that chaos can be induced in the laboratory. The first three of these researchers, for example, set up a gamelike beer production and distribution system with mock factories, wholesalers, and retailers. They placed plausible constraints on orders, inventory, and time and asked managers, trainees, and people in other roles to play the game realistically and to make reasonable business decisions. They found that these people often interacted in such a way as to produce chaos: aperiodic, unpredictable variations in inventory, huge time lags in fulfilling orders, and extreme sensitivity to small changes in conditions.

Of course, it's always dangerous and often idiotic to apply technical results outside their original domain, especially when much mathematical work remains to be done. Nevertheless, I think chaos theory (and much else) counsels that skepticism should guide us when we read about any political, economic, or military policy of any complexity. Much simpler systems governed by transparent and deterministic laws are skittishly unpredictable.

Although chaos theory casts doubt on the long-term validity of many social forecasts, it also suggests some constructive, albeit vague, ideas to keep in mind while reading stories on the economy and other social systems. One is that real change in systems often requires a reorganization of their structure. Another is that to effect change in a given system, we must search for points of maximum leverage, points that are often not obvious and are sometimes many steps removed from their intended effects. A third idea is that there is evidence that some chaos is necessary for the stability and resilience of systems.

Afta Nafta, Lafta; Free Traders Exult

Headlines and the Inverted Pyramid

The phrase *inverted pyramid* attracts the eye of a mathematician. Unbidden, the formula $1/v = 1/[^1/_3 \pi r^2 h]$ comes to mind. In journalism, of course, the notion refers to the standard practice of beginning a news story with a headline, then a lead paragraph or two in which the Who, What, When, Where, Why, and How questions are briefly answered, followed by the body of the story, the most salient details first. If a new idea is introduced later, its development has this same inverted-pyramid structure: a smaller headline, followed by a lead, and then more details. The advantage of this structure is that the story may be truncated anywhere, its end cut off for typographic or other reasons, and what remains still makes sense. A more important advantage, both good and bad, is that the reader, who generally wants to know only the bare outline of the story, can usually obtain it from the headline and the lead.

In fact, newspaper stories, with their catchy headlines, bulleted main points, and information-filled leads, are, as the journalist Jonathan Alter has noted, more likely to create a short attention span than is television, which is commonly blamed for the phenomenon. If you skim a newspaper for even fifteen minutes (unlike television, newspapers allow such random access), you can get the gist of dozens of stories. The same time devoted to watching the evening news or, worse, a TV newsmagazine, will expose you to only a few stories. The attention span required by television isn't short; it's long, but very, very shallow.

Switching stations doesn't allay your impatience with the pace of television news. When I switch between evening news broadcasts by NBC and ABC, say, I often find that they're doing the same story at the same time, especially during the first two-thirds of the program. Probability theory helps define what I mean by "often" here. Assume for the sake of illustration that only five or six stories are "big" on any given day. Assume further that the networks present these stories, all about the same length, in random order. Then a theorem on "reorderings" says that the probability that at least one of these stories will be running simultaneously on both networks at some point during the newscasts is around 63 percent. Since most would agree that some stories are inherently more newsworthy than others, the order of the stories is not random and my chances of seeing the same story on both networks at the same time are actually considerably higher than 63 percent. (The number of possible lead stories for weeklies like *Time* and *Newsweek* is greater, but even here, I suspect that chance and not collusion is the reason behind the occasional identical cover.)

Whether in television (where it's more extreme) or in newspapers, the obvious weakness of news stories' inverted-pyramid structure is that encapsulation necessarily tends to obscure complexity (more on this in a later section). And because it's so brief, the headline must rely heavily on the reader's background assumptions to make sense, and there is thus a tendency for stories to pander to popular misunderstandings or to stress conflict between ethnic groups. The availability error begs to be made. Systems and processes, whether nonlinear or not, are by their very nature not momentaveous; they don't occur at some specific moment and thus don't lend themselves to newspaper treatment unless there is crisis or breakdown of some sort. The How and Why aspects of stories about systems and processes are often particularly weak.

Happily, computers' ability to help with newspaper layouts makes typographic considerations no longer quite as constraining as they once were. News stories can sometimes be told at a pace that allows for development of the necessary historical framework and still not be continued on several inner pages. Features and human interest stories, by contrast, don't usually have an inverted-pyramid structure; they frequently begin by focusing on the intriguing predicament of some engaging individual. A compromise shape that seems to combine the need for quick communication of essential facts and a more leisurely fleshing out of these facts is a small inverted pyramid followed by a longer diamond-shaped story with a proper buildup and denouement.

As a writer of expository nonfiction, I find the accurate concision demonstrated by headlines, leads, and stories appealing and admirable. I even appreciate headlines like HEADLESS TORSO FOUND IN TOPLESS BAR. Not all headlines, however, faithfully reflect the stories under them (not to mention that the stories and the underlying reality are frequently only nodding acquaintances); witness the title of this segment. Recently, although their stories reported on the same events and appeared on the same day, the *New York Times* and *Wall Street Journal* headlined them, respectively, MEETING LAYS BARE THE ABYSS BETWEEN AIDS AND ITS CURE and SCIENTISTS HAVE AN OPTIMISTIC OUTLOOK ABOUT THE PROSPECTS IN AIDS RESEARCH.

An enlightening exercise is to read a story without seeing the headline—have somebody cut it out beforehand—decide how you would headline the piece, and then compare your headline with the real one. If that's too much trouble, switch among the evening news broadcasts and verify how frequently you find the same story running at some point during the programs.

Pakistan's Bhutto Gambles in Trade Negotiations

On Dice and Bluffing

The job of a newspaper reporter is to tell stories and to emphasize the role of the agent in any narrative. A number of segments in this book demonstrate, however, that the real story is often as dependent on chance as on intention. But what would you think of a political leader who made some important political decisions by a flip of the coin or a roll of the dice? Irresponsible? Superstitious? Nihilistic? How about rational? It's an intriguing consequence of the mathematical discipline of game theory that the conscious randomizing of choices can, if done right, maximize one's effectiveness.

In sports, the critical element of bluff is easier to see, so let me describe a typical baseball scenario that lends itself to such strategies. (The baseball example is not altogether capricious. As many have noted, baseball seems to bear the same relation to the sequential, analytic thought fostered by newspapers as football does to the immediate, visceral emotion fostered by television.) A pitcher and a batter are facing each other. The pitcher can throw either a fast ball or a curve ball. If the batter is prepared for a fast ball, he averages .500 against such pitches (that is, gets a hit 50 percent of the time) but, thus prepared, he only averages .100 against a curve ball. If he's prepared for a curve ball, however, the batter averages .400 against them, but he only averages .200 against fast balls in this case.

Based on these probabilities, the pitcher must decide which pitch to throw and the batter must anticipate this decision and prepare accordingly. If the batter prepares for a curve ball, he can certainly avoid a .100 batting average. If he always does this, however, the pitcher will throw nothing but fast balls and hold him to a .200 batting average. The batter might then decide to prepare for fast balls, which, if the pitcher continues to throw them, would give him a batting average of .500. After a while the pitcher would catch on and throw curve balls, which, if the batter continues to prepare for fast balls, would result in a .100 average for him. They could cycle about endlessly in this manner.

Each player needs to devise a general probabilistic strategy. The pitcher must decide what percentage of his pitches should be curve balls and fast balls and then make them *randomly* according to these percentages. The batter must likewise decide what percentage of the time he must prepare for each type of pitch and then do so *randomly* with these percentages in mind. The techniques and theorems of game theory allow each player in this game and a wide variety of other games to find the optimum strategies. It turns out that the solution to this particular idealized game is for the pitcher to throw fast balls half the time and curve balls the other half, and for the batter to prepare for fast balls one-third of the time and for curve balls the other two-thirds of the time. If they follow these optimum strategies, the hitter's batting average will be .300. If either deviates from his optimum strategy, he cedes an advantage to the other.

Clearly, there are many situations in business (labor conflicts and market battles), sports (virtually all competitive contests), and the military (war games) that can be modeled in this way. Although military or athletic terms are usually employed in such discussions, this vocabulary isn't essential; the subject might as easily be termed negotiation theory as game theory.

Now back to our political leader. Assume she's facing trade talks on a range of issues and products, for each of which there are several (not just two, as in the baseball example) different options she can choose. Her payoff is determined by her choice and that of the trade negotiator on the other side. If the leader is to maximize her advantage, she must let her position be dictated by a roll of the dice (and the value she places on each outcome). Given cultural attitudes, however, she will be very reluctant to admit this. Even after the negotiations, she can't blithely announce that she let her position on this or that significant matter be determined by a roll of the die. I confess that I find it oddly liberating that rational and irrational decision-making processes are sometimes indistinguishable.

Let me now sketch a different probabilistic situation. Suppose you are presented with a dial whose pointer you're requested to spin repeatedly. The pointer may stop on the red or the green half of the dial. Unbeknownst to you, however, the dial is constructed in such a way that 70 percent of the time the pointer stops on red and 30 percent of the time on green. Aside from this condition, the sequence of color outcomes is random. You're asked to predict each outcome before spinning and are rewarded according to the number of correct guesses you make during a period of up to one thousand spins. What would your response be?

Surprisingly, most people act as if they believe they have some insight into the spinner's patterns, and vary their predictions so that they're approximately 70 percent red and 30 percent green. By following this strategy, however, they guess correctly only 58 percent of the time. To see this, one needs the very important principle that the probability of several independent events all occurring is the product of these events' respective probabilities. If people guess red 70 percent of the time and the pointer stops on red 70 percent of the time, then this pair of events results in a correct guess 49 percent of the time–.7 x .7 is .49. But they also guess green 30 percent of the time, the pointer stops on green 30 percent of the time, and so this pair of events results in a correct guess 9 percent of the time–.3 x .3 is .09. Adding 49 percent and 9 percent, we get the 58 percent rate.

These people are, in effect, treating the chance component of the experiment as if it required skill, and there is a cost to this behavior. If they simply contented themselves with noticing red's 70 percent base rate, which is not difficult to do, and then acknowledging that they had no influence over or intuition into the sequence of outcomes, they could guess successfully 70 percent of the time by always predicting red. Furthermore, because the sequence of colors is random, other probabilistic predictions can readily be made.

According to the multiplication principle I cited earlier, for example, the probability that green will come up on the next four spins is $.3^4$– less than 1 percent, or about 1 chance in 123. What is the probability that red will come up at least once on the next four spins? Four consecutive times?*

A trivial example, perhaps, but, like the others, it points up the value of acknowledging uncertainty and behaving accordingly, this time by adopting an unvarying policy. Whether the issue is trade, the environment, or health care, politicians who abjure unwarranted expressions of certitude deserve plaudits, not pillory.

---

*The probability that red will come up at least once on the next four spins is $(1 - .3^4)$, or greater than .99. The probability that red will come up four times in succession is $.7^4$, or about .24.

Clinton, Dole in Sparring Roles
Who's News and Grammar Checkers

Like news content, the frequency of reporting on various newsmakers also has an inverted-pyramid structure. Important newsmakers make vastly more news than do unimportant ones. This seemingly vacuous statement suggests that we identify important newsmakers with frequent newsmakers. So be it.

Who are the principal whos about whom we read and whose whims whet our wanton inquiries into what, when, where, why, and whatnot? (Sorry.) On the front page and in the first section of a newspaper, the number-one newsmaker, the Who that's on first, is undoubtedly the president of the United States. Also big newsmakers are presidential candidates, congressional representatives, and other federal officials. Herbert Gans writes in *Deciding What's News* that 80 percent of the domestic news stories on television network news concern these four classes of people; most of the remaining 20 percent of domestic stories on the evening news cover the other 260 million of us. Gans found that fewer than 10 percent of all stories were about abstractions, objects, or systems.

Newspapers generally have broader coverage, although one study found that almost 50 percent of the sources for national and foreign news stories on the front pages of the *New York Times* and *Washington Post* were officials of the U.S. government. A large majority of those referred to on the front pages, in whatever context, were also men. Women are increasingly visible, however, receiving, according to one watchdog group, 25 percent of all references in 1994, up sharply from 11 percent in 1989.

And what about foreigners? The frequency of reporting on their Whos has the same inverted-pyramid structure; it's just that from our perspective the pyramids are much smaller. We hear from heads of state, from leaders of opposition parties or forces, and occasionally from others. The masses are not a presence at all. The journalistic rule that one American equals five Englishmen equals 500 Ecuadoreans equals 50,000 Rwandans varies with time and circumstance, but it does contain an undeniable truth: Americans, like everybody else, care much less about some parts of the world than about others. Consequently, we have no correspondents in many regions of the world, a situation that local political leaders often encourage. After all, CNN can only make these leaders' rule more precarious.

Another reason for the inverted-pyramid structure of newsmakers is that reporters naturally gravitate to where news is made, and on the federal level that place is Washington, D.C. On the state and local levels, the news comes from the state legislature and City Hall, respectively. Business, being decentralized, is largely invisible. (This is why the $12 million salary of the head of Equitable Life Insurance, for example, is seldom mentioned, even though it just about covers the salary of the entire U.S. Senate–100 salaries at $138,000 each.) More conspicuous are those businesses synonymous with their locales: Wall Street, Hollywood, and Detroit. No longer synonymous with beer, my hometown of Milwaukee almost never makes the national news unless a particularly grisly crime or some natural disaster has taken place there; the same holds for most other American cities.

The inverted-pyramid structure transcends journalistic practice. There is such a structure in linguistics as well, for example. In a large chunk of English text, the word *the* appears most frequently and is thus said to have rank order 1; the rank orders of the words *of, and,* and *to* are 2, 3, and 4, respectively. Zipf's Law states that the frequency of a word in a written document is inversely proportional to the rank order of that word. More specifically, Zipf noted that the frequency, f, of any word is the reciprocal of the product of its rank order, r, and the logarithm of the number of words, W, in the language. That is, $f = 1/[r \times \log(W)]$. Laws of this sort, which arise in a number of other human endeavors, can be used, for example, to distinguish music from sounds that are either too random or too ordered (corresponding to an exponent of r other than one).

I know of no studies that investigate the question of a Zipf-like law relating the frequency of occurrence of a particular newsmaker and his or her rank order of occurrence, but it's not unreasonable to conjecture that there is such a rough relationship. If there is, the president would be the analogue of the word *the* in American newspapers. Perhaps there is also such a formula describing the frequency of occurrence of foreign countries in American newspapers, in which case the adage about the relative coverage of Americans, Englishmen, and other nationalities might also be quantified. Interestingly, if these Zipf-like laws are graphed on a logarithmic scale, we obtain downward-slanting straight lines suggesting the shape of an inverted pyramid! There is no great insight here, but it is satisfying when mathematics corroborates our commonsense observations, in this case about newsmakers and inverted pyramids.

Let me close with a little experiment I performed that is germane to the issue of newspapers and linguistics. Many word-processing programs come equipped with grammar-checking utilities. The only mistakes they're ("Are you sure you don't mean their?") likely to catch are ones so glaring that the person making them needs more help than this software can provide. Nevertheless, they're interesting to play with. ("Are you sure you want to end a sentence with a preposition?") They print out a variety of statistics on any passage they analyze—number of letters per word, number of words per sentence, number of sentences per paragraph, percentage of prepositions, frequency of words used, frequency of the passive voice, and so on.

Despite their rudimentary nature, one might find these grammar checkers useful as aids in matching the style of a given piece of writing. This is what I did. After determining the grammatical characteristics of a few tabloid news stories (whose simple prose is relatively easy to simulate), I wrote a story that syntactically conformed to them. Paying no particular attention to content, I merely strung together some customary newspaper locutions, a few familiar characters and phrases, and gushing expressions, all with the appropriate frequency and length. If read very rapidly, the piece sounded tabloidy enough. I headlined it TORNADO KILLS FIVE, SELF. Using more powerful and intelligent software, spending more time on the editing, and insisting on a bit of superficial coherence, one could easily approach the sound (but not the sense) of more substantive pieces.

Iraqi Death Toll Unknown

Benchmark Figures in War, Crisis, and the Economy

News stories are often bereft of the numbers that would enable a reader to put them in perspective. This is particularly true in times of war when perspective is in short supply.

I recall, for example, various estimates of the number of Iraqi military casualties during the Gulf War. Under vague headlines there appeared dispatches with few numbers and scant analysis. Little effort was expended to come up with the casualty figures, but most present estimates range between 40,000 and 80,000 Iraqi soldiers killed and an indeterminate number wounded. One needn't (and I don't) feel any sympathy for Saddam Hussein or his thuggish Baathist regime to be shocked by these numbers. Iraq is a country of 18 million people, approximately 7 percent of the U.S. population. A comparable loss of American soldiers would have resulted in 570,000 to 1,140,000 dead, the latter figure being twice the number of our total forces then in the entire region. Surely these numbers, which during the war were whispered to be much higher, could have been reformulated and put in perspective by some of the horde of reporters covering the war. Such reformulations would have been more informative than the number of sorties flown, a precise but relatively meaningless figure that was recited, mantralike, almost hourly. The so-called fog of war explains some of the poor coverage (the ineffectiveness of the Patriot missiles in defending Israel is a prime instance), but certainly not all of it.

Likewise, the approximate number of Vietnamese who died during the Vietnam War was rarely guessed at in print and remains largely unappreciated. While 58,000 Americans perished, an estimated 2–2.5 million Vietnamese (out of a population of about 65 million) died. The exact figure is unknown, and perhaps even unknowable. Americans missing in action (M.I.A.'s) from that war number approximately 2,000, compared with an estimated 200,000 missing Vietnamese. By way of comparison, there were approximately 8,000 American M.I.A.'s in Korea and close to 80,000 in World War II. Somehow the M.I.A. issue never became politicized in those wars, however.

The issue of benchmark (or ballpark) figures transcends war. In smaller, more civil conflicts, for example, agreement on some basic numbers can sometimes provide a common ground. In fact, I think such figures assume a greater importance in multicultural societies where a common culture, stories, and myths are not as widely shared as they are in more homogeneous societies. In any case, such figures should be a part of major stories or should at least appear periodically in the ongoing coverage.

Consider the many news stories on the speeches of Khalid Abdul Muhammad. His and others' claim that 600 million African-American deaths were attributable to slavery went unchallenged, even though encyclopedias put the total number of slaves brought to the New World at somewhere between 8 million and 15 million. Similar elementary research could have demolished the absurd assertion by Nation of Islam leader Louis Farrakhan that 75 percent of American slaves were owned by Jews. At the beginning of the Civil War, Jews constituted 0.22 percent of all Southerners (20,000 of 9 million) and 0.26 percent of slaveholders (5,000 of 1.9 million).

These considerations apply as well to more mundane stories. If the *Chicago Tribune* were to report, for example, that two piano tuners died under mysterious circumstances in the past year, we would need to know the approximate number of piano tuners in the city to evaluate the possible significance of this fact. What are some plausible assumptions necessary to come up with a reasonable estimate for this number?*

The downside to the inclusion of round figures in stories is that they can become stale and immune to revision. *Spy* magazine did a short feature on the statistic that one million Americans are infected with the AIDS virus. It cited stories from major newspapers dating from 1985 to 1993, all mentioning this big round number. In March 1994, the *New York Times* acknowledged the oddly static nature of this number in an article on the difficulties of precisely ascertaining it: long incubation period, privacy considerations, changing definitions.

Reading a generic headline, MANY MORE CASES EXPECTED, about any malady or condition should trigger a number of natural questions. If I may pose a meta-question, what are some of them?†

Articles on economics can benefit particularly from the occasional inclusion of appropriate benchmark estimates (although here, too, it can be overdone). If the subject is the national economy, for example, then mentioning the fact that the annual GNP is approximately $6 trillion serves to orient the reader. Knowing that the budget is about $1.5 trillion, that entitlements constitute more than half the budget, that the national debt is almost $4 trillion, that the budget deficit is $0.3 trillion, or $300 billion, can be immeasurably helpful.

Furthermore, these numbers should, if possible, be compared with quantities that are more viscerally appreciated. For example, estimates for the cost of the savings-and-loan bailout have ranged up to $500 billion (including interest payments over time). This translates into $2,000 for every man, woman, and child in the United States (again, over time). It would also pay for approximately 12 mil-

---

*The following questions would need to be answered: How many people are there in Chicago? How many households? What fraction of them have pianos? How many schools are there? How frequently is the average piano tuned? How many pianos are tuned in one week by an average piano tuner? An order-of-magnitude estimate for these numbers is all that's required, not a trip to the library.

†Some natural questions might be: What is a "case" and what is not? More precisely, how many more are expected? How many have there been already? How many is normal? Who is doing the expecting? Might there be an element of self-fulfilling prophecy or a hidden agenda? Are all cases being counted? What about their rate of increase?

lion Mercedes, 20 million Volvos, 30 million Mazdas, or 100 million Eastern European cars. Or, given that gold sells for roughly $350 an ounce and that the distance from the East Coast to the West Coast is about 15 million feet, $500 billion could buy a transcontinental gold bar weighing about 5-1/2 pounds a foot (ignoring the price rise that would result from this goldbricking project). If we were to stretch this gold bar into a rainbow extending from Capitol Hill 1,500 miles up, above the Midwestern prairies and over the Phoenix headquarters of Charles Keating's failed savings-and-loan empire, this golden arch would weigh in at almost 4 pounds a foot. It would take a decade to spend $500 billion at $1,585 a second, and if spending at this rate occurred only during business hours, more than forty years would be required to use it up.

For a more somber example, the United Nations Children's Fund (UNICEF) reports that millions of children die each year from nothing more serious than measles, tetanus, respiratory infections, and diarrhea. The latter two maladies killed, respectively, eight and six times as many people worldwide as AIDS did in 1990, according to the Harvard School of Public Health. These illnesses can be prevented by a $1.50 vaccine, $1.00 in antibiotics, or 10 cents' worth of oral rehydration salts. UNICEF estimates that $2.5 billion (most of it going to staff and administration) would be sufficient to keep these children alive and improve the health of countless others.

Another relatively invisible health problem that requires only sufficient will to stop is the practice of clitoridectomy. This mutilation of the female genitalia is still disgustingly prevalent in large parts of the world. (I plead guilty to cultural imperialism.) When I was in the Peace Corps in Kenya, I was invited to observe the procedure in the same way I might be invited to a confirmation or bat mitzvah here.

Of course, these equivalences beg a number of questions about comparability, but most of us tend to be mesmerized by "people" stories and bored by "number" stories like the savings and loan (S & L) scandal. Ultimately, the distinction between the two is specious. Unless we find better ways to vivify complex issues and to keep our heads in times of crisis, the cost to us will continue to be much more than $500 billion.

D'Amato Agrees Hillary Most Honest Person Clinton Knows

Ambiguity and Nonstandard Models

The pledge of "no new taxes" can easily be fulfilled if we redefine *taxes* adroitly enough. The same technique can be used to declare victory on health care legislation, welfare reform, and a myriad of other issues. Knowing when to redefine terms, when to interpret them loosely, and when to take them literally is part of the politician's craft and is necessary, as George Orwell wrote, "to give an appearance of solidity to pure wind."

In defending his wife's integrity, under attack because of allegations of preferential treatment for her commodities dealings, President Clinton stated that she was the most honest person he knew. New York's Senator Alphonse D'Amato, no stranger to controversy himself and one of the First Lady's chief critics, replied that the president had spoken the literal truth. So had the professor of logic who, when the elevator he was riding stopped and opened and he was asked if it was going up or down, answered "yes." Ambiguity, literalness, irony, redefinition—these are the elements of the artful interpretation that is the most characteristic aspect of political discourse.

One of my favorite instances is from the world of petty politics, otherwise known as academia. It is a charmingly equivocal letter of recommendation:

You write to ask me for my opinion of X, who has applied for a position in your department. I cannot recommend him too highly nor say enough good things about him. There is no other student of mine with whom I can adequately compare him. His thesis is the sort of work you don't expect to see nowadays and in it he has clearly demonstrated his complete capabilities. The amount of material he knows will surprise you. You will indeed be fortunate if you can get him to work for you.

And what does mathematics have to say about ambiguity? Well, it lends theoretical support to the idea that recognizing and avoiding ambiguity are not easy. In fact, one of the difficulties mathematicians face in devising a collection of statements that characterizes some mathematical entity, such as the set of whole numbers, is ruling out alternative interpretations. Logicians call these alternative interpretations *nonstandard models*. For example, mathematicians might propose that the whole numbers be considered to be any set of objects with operations $+$ and $\times$ defined on them that satisfy certain axioms such as $X + Y = Y + X$, $X(Y + Z) = (X \times Y) + (X \times Z)$, and so on. Later, they may be surprised to discover that all sorts of other things that aren't at all what they mean by numbers also satisfy these axioms. So-called non-Euclidean geometries, useful in fields as diverse as navigation and relativity theory, can be thought of as unexpected nonstandard models of (some of) Euclid's axioms.

Nonstandard models and interpretations play a role not only in politics and mathematics but in humor as well. An old chestnut that is appropriate in a book about newspapers is the joke "What's black and white and re(a)d all over?" Wounded zebras are a well-known nonstandard model. Another concerns the Texan who was boasting to an Israeli kibbutznik about the size of his ranch and remarked that it took him all day to drive across it in his jeep. The kibbutznik rejoined that he had a jeep like that once, too. And then there was the man who answered a matchmaking company's computerized personals ad in the paper. He expressed his desire for a partner who enjoys company, is comfortable in formal wear, likes winter sports, and is very short. The company matched him with a penguin.

If pinning down notions, even precise mathematical ones, is so difficult, and humor is frequently the result of failing to do so, then it's perhaps not astonishing that politics and humor are so inextricably bound. It's no accident that sobriquets like Tricky Dick, Teflon Ron, and Slick Willie regularly attach themselves to the nation's premier politicians.

Fraud Alleged in Pennsylvania Senate Race

Political and Mathematical Regression

The *Philadelphia Inquirer* has reported extensively on what appears to have been large-scale fraud in Pennsylvania's Second State Senatorial District in Philadelphia, a district crucial to the balance of power in the state legislature. The fraud involved absentee ballots in the recent special election. Although the Republican candidate received more votes on the voting machines than did the Democratic candidate, the latter received an overwhelming number of absentee ballots, allowing him to eke out a narrow victory.

The authenticity of many of the absentee ballots, however, was in question. The case ended up in court, where mathematics experts for the judge, the Democrats, and the Republicans all offered different slants on the election results. (Having been involved in a couple of lawsuits as an expert probability witness and having observed that a prudent skepticism is often less prized than an indefensible certainty, I turned down preliminary requests from both sides to testify.)

The expert hired by the judge took the following approach. He examined the records of twenty-two previous elections in Philadelphia's senatorial districts and recorded the difference between the Democratic vote and the Republican vote as registered by voting machines. Next he recorded the difference between the Democratic vote and the Republican vote as registered by absentee ballots in all these elections. Then for each election, he plotted these pairs of numbers on a graph and determined the best straight line through these points (see diagram). Most of the points are clustered near this line, known in statistics as the *regression line*. It gives the approximate relation between the absentee vote and the machine vote. If these past elections are a reliable predictor, the chances of such a large discrepancy between absentee and machine results as that which occurred in the disputed election were determined to be very low, about 6 percent.

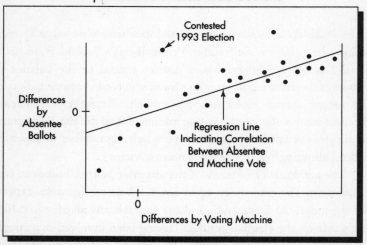

Differences Between Democratic and Republican Vote Totals in 22 Elections

The expert hired by the Democrats pounced on this "if." He argued that past elections were not necessarily an accurate guide and speculated that the Democrats' zeal in attracting absentee votes in this particular election, rather than fraud, might have accounted for the discrepancy. The expert for the Republicans concentrated on the considerable evidence that most of the absentee ballots had been illegally cast. If these ballots were to be thrown out, he continued, the Republican candidate would be a shoo-in. (There are other interpretations of the figures as well.) The upshot: The judge seated the Republican.

Another recent case in which mathematics buttressed varying interpretations was the Dinkins-Giuliani mayoral election in New York City, mentioned in the introduction. Claims were made that blacks vote along racial lines more than whites do. The evidence cited was that 95 percent of blacks voted for (black) mayor David Dinkins, whereas only 75 percent of whites voted for (white) candidate (and victor) Rudolph Giuliani. This failed to take into account, however, the preference of most black voters for *any* Democratic candidate. Assuming that 80 percent of blacks usually vote for Democrats and only 50 percent of whites usually vote for Republicans, one can argue that only 15 percent of blacks voted for Democrat Dinkins based on race, but 25 percent of the whites voted for Republican Giuliani based on race. Again, there are other explanations.

These cases illustrate the way we use mathematics but are not necessarily bound by it. Consider a more schematic illustration. Two men bet on a series of coin flips, agreeing that the first to win six flips will be awarded the $64,000 stake. The game, however, is interrupted after only eight flips, with the first man leading five to three. The $64,000 question is, How should the pot be divided? The first man might be expected to argue that he should be awarded the full amount because the bet was all or nothing and he was leading. The second man would probably argue that the bet was off or that they should split the pot since the game wasn't finished.

A more neutral party might reason instead that the first man should receive $5/8$ of the money ($40,000) and the other the remaining $3/8$ ($24,000) because the score was five to three. Another disinterested observer might contend that because the probability of the first man's going on to win can be computed to be $7/8$ (the only way the second man can win is by getting three flips in a row, a feat with only a $1/8$ probability: $1/2 \times 1/2 \times 1/2$), the first man should receive $7/8$ of the pot, leaving only $8,000 for the second man. This, incidentally, was Pascal's solution to this, one of the first problems in probability theory. Rationales for still other divisions are possible. Can you think of any?

The point is that the criteria for deciding on any one of these divisions are nonmathematical. Mathematics can help determine the consequences of our assumptions and values, but we, not some mathematical divinity, are the origin of these assumptions and values. That's why we have judges and legislatures, even if their actions are sometimes less than pristine.

Cult Members Accuse Government of Plot

Newspapers, Coincidences, and Conspiracy Theories

Turning the pages of a newspaper or flipping through the channels on TV can be disorienting. Everything seems to be of equal importance, especially if one is unfamiliar with previous developments in ongoing stories. Part of the fascination of newspapers, this unnatural cross section of times and locations is certain to appear disorderly and confusing even when the underlying phenomena are not. But there is also a certain illusion of power and invulnerability one gets by the chance sampling of disparate stories in a paper—a murder here, a hostile takeover there, a sad story of war juxtaposed with a loss by the local baseball team (sometimes, oddly, more affecting), a new health finding next to a dismal story on the decline in educational standards.

A consequence of these feelings is an easy receptivity to the meaningless coincidences and incongruities that permeate any newspaper. These have always constituted a mother lode of material for standup comedians, who merely have to skim the paper in order for odd juxtapositions to jump out at them. Lorena Bobbitt severs her husband's penis and, at about the same time, former House Speaker Thomas "Tip" O'Neill dies. The happenstance results in the local headline ALL BOSTON MOURNS "TIP," suggesting inordinate sympathy for Mr. Bobbitt in Beantown.

There are so many ways in which events, organizations, and we ourselves may be linked that it's almost impossible to believe in the significance of all of them. Yet many do, sometimes arguing that the probability of this or that coincidence is so low that it must mean something. Such people fail to realize that though it is unlikely that any *particular* sequence of events specified beforehand will occur, there is a high probability that *some* remarkable sequence of events will be observed subsequently. This is especially so when one is inundated with so much decontextualized information.

If we actively set out to look for coincidences in the daily paper, we could cross-check key words, phrases, names, and numbers by computer and come up with hundreds of them.* If we don't specify the particular nature of the connections beforehand, we can rely on the vast number of possible relations to lead to innumerable suggestive associations. In the throes of obsession, the conspiracy theorist searches not for arbitrary coincidences but only for those that support his beliefs—and because of the myriad connections among items in the paper, he is almost always successful. The absence of certain items in the paper is also conducive to such theorizing, and there's no telling how to count such absences. There is always enough evidence out there, positive or negative, for people who want or need to believe badly enough.

Historical and quasi-historical examples abound as well. A friend of mine who is a rabbi notes that the Gulf War ended on the Jewish holiday of Purim, which celebrates the defeat of the evil Haman by Mordecai. Moreover, behold that on this day millennia later and in the same part of the world, the dictator Hussein, whose name also begins with an "H," is vanquished by a man whose name also begins with an "M"–Major General Schwarzkopf! He intended this as a kind of joke, but fanatics of all stripes have made much of even flimsier links.

This process brings to mind the list of well-known connections

---

*The branch of mathematical combinatorics known as Ramsey theory is relevant here. It studies how big sets must be so that certain relations are guaranteed to hold among some of their elements. How many guests need be present at a party, for example, to ensure that at least five of them all know one another or are strangers?

between Presidents Lincoln and Kennedy. Lincoln was elected president in 1860, Kennedy in 1960. Their names both consist of seven letters. Lincoln had a secretary named Kennedy and Kennedy had one named Lincoln. Lincoln and Kennedy were assassinated by John Wilkes Booth and (allegedly) Lee Harvey Oswald, respectively, men who went by three names and who advocated unpopular political positions. Booth shot Lincoln in a theater and fled to a warehouse; Oswald shot Kennedy from a warehouse and fled to a theater.

John Leavy, a computer programmer at the University of Texas, wondered whether similar lists might be constructed for *any* pair of U.S. presidents. To test his conjecture, he fed data on the presidents into a computer and came up with correspondences between pairs of presidents that were just as remarkable–and hence just as unremarkable–as those between Lincoln and Kennedy. One of the examples he published in *The Skeptical Inquirer* concerns two other assassinated presidents, William McKinley and James Garfield.

It turns out that both of these presidents were Republicans who were born and bred in Ohio. They were both Civil War veterans, and both served in the House of Representatives. Both were ardent supporters of protective tariffs and the gold standard, and both of their last names contained eight letters. After their assassinations they were replaced by their vice presidents, Theodore Roosevelt and Chester Alan Arthur, who were both from New York City, who both sported mustaches, and who both had names containing seventeen letters. Both presidents were slain during the first September of their respective terms by assassins, Charles Guiteau and Leon Czolgosz, who had foreign-sounding names. Not being superstars of American history, McKinley and Garfield do not attract the same intense fascination as do Lincoln and Kennedy, however.

Exercise: Dream up your own coincidence-inspired theory and support it with as much circumstantial and adventitious evidence as you can find. If you're particularly successful in your research, you may want to submit your creation to *The National Inquirer*.

Doing a few searches through Nexis and related newspaper and commercial databases and seeing the ease with which coincidentally generated conspiracy theories can be constructed greatly undermines their appeal. Coincidences are sometimes significant, of course, and occasionally there are real conspiracies. But most coincidences are meaningless, and most conspiracies products of feverish fantasies. I suspect that relatively few real conspiracies remain undiscovered for long; people like to talk.

Section 2

LOCAL, BUSINESS, AND SOCIAL ISSUES

This book presents what all the people in the world are doing, at the same time, in the course of one minute.

—STANISLAW LEM

Although the grouping together of news articles can be somewhat arbitrary, many newspapers commonly carry local and business news as well as stories on social issues in their second sections. Thus the segments in this section deal with questions like the following: Are statistical disparities sure signs of racism? Is the SAT predictive of college success? Are abortion opponents susceptible to a reductio ad absurdum argument? And what about guns, cars, and the converging death rates resulting from each? Why should defense lawyers stress the difference between a conditional statement and its converse? How much do stockbrokers in lavish suites differ from storefront palm readers? Should we be worried about cellular phones that cause brain cancer or trucks that explode? Can arithmetic be employed to help sell products? Can novelty be quantified? And what is truly local nowadays?

A good place to start is with the last question. Certainly a disproportionate share of what's termed local news is devoted to reporting crime, malfeasance, and other regional miseries. In a smaller population, such calamities are not generally as extreme or heinous as are national or international ones and thus impart a feeling of relative safety: Really bad stuff happens elsewhere; we're all above average around here. Most of us become accustomed to the standard stories on drug use, school problems, strikes, embezzlement, civic corruption, and the like in our communities. We come to expect a certain rate of homicides, automobile accidents, and disease. These rates, we also learn, vary significantly from neighborhood to neighborhood. Most murders in Philadelphia, according to a recent article in the *Philadelphia Inquirer*, are clustered in three or four districts in the city. AIDS cases in New York City occur throughout the city but are vastly more common occurrences within just a few ZIP codes.

Reading all these stories tends to inure us to the tragedies underlying them; we lose our sense of connection. We learn of the time, location, or circumstances of some misfortune, realize we would never be in that place at that time or in that situation, and go on to another part of the paper. Unfeeling, perhaps, this attitude is also quite necessary if we're not to dissolve into a quivering blob of indiscriminate compassion. We can't respond to all these disasters, and so we pick and choose and develop scales of importance that are not always tied to proximity (or to anything else, for that matter).

Geographically local becomes less important to us; what is economically, culturally, and environmentally local becomes more so. People in an upper-middle-class suburb of Philadelphia are generally more sympathetic to their similarly situated fellows outside Chicago than they are to inner-city Philadelphians, who might, in turn, be more understanding of the residents of a comparable district in Houston. Researchers at a university physics department in Arizona will probably communicate (via E-mail) more frequently with their counterparts in Massachusetts than with their colleagues in the English department across campus. People with the same ethnic background often feel a more immediate connection to each other than they do to their neighbors down the road.

This expansion of what is local doesn't apply to people alone. The smokestacks in the Midwest are "close" to the lakes in eastern Canada. Chernobyl is "close" to Stockholm. Hollywood's films are ubiquitous. CNN brings the world to us. A fad is born and almost instantaneously adopted half a continent away. And the headline PORNOGRAPHY SUBVERTS COMMUNITY STANDARDS is borne out by computer bulletin boards that dispense pictures and stories that are of world-class "artistry."

Does mathematics have anything to add to these truisms? The

so-called multiplication principle is relevant. It states that if one has M choices to make and each leads to N other choices, then there are M × N possible pairs of choices one can make. And if each of these M × N pairs of choices leads to P further choices, then there are M × N × P possible three-step sequences of choices one can make, and so on. Thus, there are 26^3 different sets of three initials. Likewise, the number of different telephone numbers within an area code is no more than eight million: eight possible digits for the first position (0 calls the operator and 1 leads to the long-distance system), followed by ten possible digits for each of the second, third, fourth, fifth, sixth, and seventh positions, for a total of 8×10^6. There are other constraints on possible numbers, but I'll ignore them here and merely note that the argument indicates why area codes are necessary. By the way, given phone company constraints and the multiplication principle, how many area codes are possible?*

Using the multiplication principle and certain reasonable empirical assumptions, we can show that two randomly selected American adults are linked by two intermediates almost 99 percent of the time. (Oscar and Myrtle are linked in such a way if there exist other people X and Y such that Oscar knows X who knows Y who knows Myrtle. *Know* must be defined, of course, and Oscar and Myrtle needn't be aware of their connection.) The numbers will differ, but comparable results hold for citizens of the world. The lesson's the same: Local is not what it used to be, and we shouldn't be surprised at how closely we're linked.

This idea of the number of links between people can be explored further. In a well-known experiment, the psychologist Stanley Milgram randomly selected pairs of American adults, designating one of them the source and the other the target. He gave each source the name and address of a target and directed the source to send a letter to the person in the country who, he or she judged, was most likely

---

*The first digit of an area code can be anything except 0 or 1, the second must be 0 or 1, the last can be anything. Thus there are $8 \times 2 \times 10 = 160$ area codes possible. Since fax machines, cellular phones, and several lines to the same home or business are rapidly filling the available codes, the requirement that 0 or 1 appear in the second position may soon be relaxed.

to know the target. He also directed that a cover letter be included directing the recipient of the source's letter to do the same thing. This was to continue until the target was reached. (If the target lived in Seattle, for example, the source might have sent the letter to an acquaintance in Portland, who might have forwarded it to someone in Tacoma who might, in turn, have had a brother-in-law who...) Milgram found that the number of intermediate links between source and target ranged from two to ten, with five being the most common number. This empirical study is even more telling than the a priori probability argument about the mere existence of two intermediates and partially explains why privy information, gossip, and jokes cascade so rapidly through a population. The number of links would be even smaller if further information about the target were provided to the source.

It's not clear how one would carry out studies to confirm this, but I suspect that the average number of links connecting an arbitrary pair of people has shrunk over the last fifty years. Furthermore, this number will continue to shrink because of advances in communication and despite an increasing population. (Of course, the definition of link is crucial and must be fairly liberal for this conjecture to be true.) One consequence is that we're indirectly exposed to more mayhem and disaster than ever before, contributing to our above-mentioned numb response to them.

Often, a feeling for an organization (company, town, school) can be obtained by drawing a graph or network tracing the professional and personal links among its members. These local organizations are not islands, however, and a more ambitious graph linking them to larger entities and their members along even more dimensions can at least be imagined, if not actually drawn. Continue expanding this graph and, as a last stage in this thought experiment, imagine all 5.6 billion of us and the links that connect us to one another in so many different ways: job, profession, interests, geography, physical type, religion, and so on. We will be aware of detail in our own local organizations, and there will be discernible coherence in more diverse groupings, but most of this profusely spaghettied graph will forever remain invisible to us.

We might nevertheless be interested in defining especially active nodes in this network, reasoning that they are powerful or important in some other sense. Toward this end, imagine measuring the frequency of communication between any two nodes and counting the number of different ways they can be linked via one- or two-step paths. A node linked via many busy one- or two-step paths to millions of others may be deemed significant. New communication modalities such as C-Span or the Internet introduce new paths through the graph, as do organizational realignments. I find this huge graph and its myriad subgraphs to be a useful metaphor with which to think about our increasingly intertwined world.

Consider, for example, the sex subgraph that connects two people, say George and Martha, if there is a set of intermediates such that George has had sex with A who has had sex with B who has had sex with C and so on, until Martha is reached. If we place all people who are connected to one other in this way into the same group, this relation divides all Americans into nonoverlapping groups of people. My guess is that there are celibates who are in their own single-person groups; a large number of monogamous dyads neither of whose members ever had sex with anyone else and thus constitute their own two-person groups; relatively few groups having three, four, five, or a smaller number of members; and then the rest of the U.S. adult population in one large group containing 100 million or so members. The vast majority of the latter group is not promiscuous. The huge size of the group derives from our interconnectedness.

And where does the mundane newspaper fit into this? Its job may be ideally conceived as covering and magnifying the main nodes and linking thoroughfares of a section of this unimaginably complex web that binds us all. This may be a bizarre notion of a newspaper, but it serves to counter the equally bizarre conceptions of the paper as a sort of glorified police blotter or merely a tool of this or that elite.

Returning to the metropolitan newsroom from which I began this convoluted journey, I observe that which local nodes and thoroughfares are to be covered is a matter of strict journalistic convention (tempered by courage, vested interest, and passion). As in the case of national news, which focuses almost exclusively on the presidency and the official organs of government, conventional local news focuses inordinately on certain common venues. As mentioned, there is the City Hall beat, fixing on the mayor and the various local governmental offices. The Board of Education and the Police Department are also at the crossroads of many different public paths, and reporters can usually get a story by simply standing by. Local business is neither public nor centralized, so there is considerably less newspaper coverage of companies except during times of labor unrest. Profiles of local personalities, from a variety of domains but lopsidedly from the local media itself, also appear. But when it comes to geographically local news, the number of answers to the journalist's Where is disappointingly small—usually in one of the four or five standard places.

Company Charged with Ethnic Bias in Hiring

Test Disparities Need Not Imply Racism

The comedian Mort Sahl remarks that some newspapers might report a nuclear exchange between the United States and Russia with the headline WORLD ENDS: WOMEN AND MINORITIES HARDEST HIT. Sarcasm and hyperbole aside, victimization and the differential treatment of groups, whether intentional or not, are the basis for many a news story. The percentage of African-American students at elite colleges, the proportion of women in managerial positions, the ratio of Hispanic representatives in legislatures have all been written about extensively. Oddly enough, the shape of normal bell-shaped statistical curves sometimes has unexpected consequences for such situations. For example, even a slight divergence between the averages of different population groups is accentuated at the extreme ends of these curves, and these extremes often receive inordinate attention in the press. There are other inferences that have been drawn from this fact, some involving social policy issues such as affirmative action and jobs programs. The issue is a charged one, and I don't wish to endorse any dubious claims, but merely to clarify some mathematical points.

As an illustration, assume that two population groups vary along some dimension—height, for example. Although it is not essential to the argument, make the further assumption that the two groups' heights vary in a normal or bell-shaped manner (see diagram). Then even if the average height of one group is only slightly greater than the average height of the other, people from the taller group will constitute a large majority among the very tall (the right tail of the curve). Likewise, people from the shorter group will constitute a large majority among the very short (the left tail of the curve). This is true even though *the bulk of the people from both groups are of roughly average stature*. Thus if group A has a mean height of 5'8" and group B a mean height of 5'7", then (depending on the exact variability of the heights) perhaps 90 percent or more of those over 6'2" will be from group A. In general, any differences between two groups will always be greatly accentuated at the extremes.

Two Normal Curves

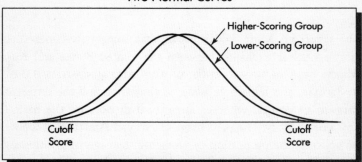

Small differences in the mean lead to
large differences at the extremes.

These simple ideas can be used and misused by people of very different political persuasions. My concerns, as I've said, are only with some mathematical aspects of a very complicated story. Let me again illustrate with a somewhat idealized case. Many people submit their job applications to a large corporation. Some of these people are Mexican and some are Korean, and the corporation uses a single test to determine which jobs to offer to whom. For whatever reasons (good or bad, justifiable or not), let's assume that although the scores of both groups are normally distributed with similar variability, those of the Mexican applicants are slightly lower on average than those of the Korean applicants.

The corporation's personnel officer notes the relatively small differences between the groups' means and observes with satisfaction that the many mid-level positions are occupied by both Mexicans and Koreans. She is puzzled, however, by the preponderance of Koreans assigned to the relatively few top jobs, those requiring an exceedingly high score on the qualifying test. The personnel officer does further research and discovers that most holders of the comparably few bottom jobs, assigned to applicants because of their very low scores on the qualifying test, are Mexican. She may suspect racism, but the result might just as well be an unforeseen consequence of the way the normal distribution works. Paradoxically, if she *lowers* the threshold for entrance to the mid-level jobs, she will actually end up *increasing* the percentage of Mexicans in the bottom category.

The fact is that groups differ in history, interests, and cultural values and along a whole host of other dimensions (which are impossible to disentangle). These differences constitute the group's identity and are what makes it possible even to talk about a collection of people as a group. Confronted with these social and historical dissimilarities, then, we shouldn't be astonished that members' scores on some standardized test are also likely to differ in the mean and, much more substantially, at the extremes of the test-score distribution. (Much of this discussion is valid even if the distribution is not the normal bell-shaped one.) Such statistical disparities are not necessarily evidence of racism or ethnic prejudice, although, without a doubt, they sometimes are. One can and should debate whether the tests in question are appropriate for the purpose at hand, but one shouldn't be surprised when normal curves behave normally. As long as I'm issuing pronouncements, let me make another: The basic unit upon which our society or, indeed, any liberal society ("indeed" is a sure sign of something pompous coming up) is founded is the individual, not the group; I think it should stay that way.

Aside from having a questionable rationale, schemes of strict proportional representation are impossible to implement. Another thought experiment illustrates this point. Imagine a company—let's call it PC Industries—operating in a community that is 25 percent black, 75 percent white, 5 percent homosexual, and 95 percent heterosexual. Unknown to PCI and the community is the fact that only 2 percent of the blacks are homosexual, whereas 6 percent of the whites are. Making a concerted attempt to assemble a workforce of 1,000 that "fairly" reflects the community, the company hires 750 whites and 250 blacks. However, just 5 of the blacks (or 2 percent) would be homosexual, whereas 45 of the whites (or 6 percent) would be (totaling 50, 5 percent of all workers). Despite these efforts, the company could still be accused by its black employees of being homophobic, since only 2 percent of the black employees would be homosexual, not the communitywide 5 percent. The company's homosexual employees could likewise claim that the company was racist, since only 10 percent of their members would be black, not the communitywide 25 percent. White heterosexuals would certainly make similar complaints.

To complete the reductio ad absurdum, factor in several other groups: Hispanics, women, Norwegians, even. Their memberships will likely also intersect to various unknown degrees.* People will identify with varying intensity with the various groups to which they belong (whose very definitions are vague at best). The backgrounds and training across these various cross sections and intersections are extremely unlikely to be uniform. Statistical disparities will necessarily result.

Racism and homophobia and all other forms of group hatreds are real enough without making them our unthinking first inference when confronted with such disparities.

---

*A partially rhetorical question of tangential relevance: Assume that an organization wishes to "encourage" those having some characteristic C, but cannot directly inquire of anyone whether he or she possesses it. Assume further that Mr. X has a surname 20 percent of whose co-owners have characteristic C. If one knows nothing else about Mr. X, then it seems prudent to suppose that there is a 20 percent chance that Mr. X possesses C. If one later discovers that Mr. X comes from a neighborhood 70 percent of whose members have characteristic C, what should one's estimate now be of the likelihood that Mr. X posesses C? And what if one subsequently learns that Mr. X is an active member of a nationwide organization only 3 percent of whose members possess characteristic C? With all this information, what can one now conclude about the chances that Mr. X has C?

SAT Top Quartile Score Declines
Correlation, Prediction, and Improvement

The Scholastic Assessment Test (SAT) is familiar to anyone who's ever attended high school in this country. Retaining the brand-name acronym, its new name is slightly less presuming than the former Scholastic Aptitude Test. Not surprisingly, some of the hundreds of news stories on the SAT spotlight those rare students who get a perfect 800 on the math section or a perfect 1600 on the test as a whole, much less frequently those who score 800 in the English section only. The standard newspaper filler describes the local student's high school activities, admirable qualities, college and career goals, and usually some endearing foible.

But most pieces are more substantive. A check of various newspaper databases reveals that in recent years, stories have dealt with the decline of scores since the early 1960s, the possible addition of essay questions to the heretofore exclusively multiple-choice format, the permission to use calculators, the poorer performance of girls and minorities on the test, the "renorming" of the test to inflate the average score to 1000, the extent to which scores can be raised by services that focus on test-taking strategies and heuristics, and, most important, the ability of the test to predict collegial success.

An amazing aspect of the ongoing coverage has been its breadth, detail, and tendentious slant. For what they're worth, my brief reactions to the issues just mentioned: Scores have declined in part because a larger and more varied body of students is taking the test than ever before. Unfortunately, the scores in the top quartile have also declined, indicating a more general deterioration. The inclusion of essays on future tests would be a welcome modification. It doesn't require a score of 1600 to realize that the ability to organize and express one's thoughts in a coherent essay without misspellings or grammatical errors is at least as important as other skills the test measures.

Calculators will be of marginal help to test takers; the difficulty students have with the math questions is, popular opinion to the contrary, not computational. Girls consistently score lower than boys (only one-third of the highest scorers on the PSAT are girls), and minorities consistently score lower scores than whites. Nevertheless, the frequent charges of cultural and gender bias are, in my opinion, overstated (despite the rare question of the runner-is-to-marathon as oarsman-is-to-regatta type). The test is "biased," but only toward the educationally prepared, the physically healthy, and the psychologically receptive. And the gain from renorming the scores upward to achieve a symmetric distribution ranging between 200 and 1600 is outweighed by two factors: the loss of discriminatory power at the upper end of the distribution and the lack of easy comparability between future and past scores. Why not adjust the liveliness of baseballs so that the average hitter bats .500?

Studies do suggest that attending SAT preparation courses will raise one's scores to some degree. (More extensive and long-term intervention will raise them even more, of course.) Nevertheless, if students have sufficient self-discipline to practice on the sample tests supplied by the Educational Testing Service or to work with inexpensive software, I think they can derive a good deal of the benefit provided by these services on their own. (One striking hint some services offer is to learn how to darken circles quickly. Taking a second or two longer than is necessary on each answer can conceivably result in a noticeably lower score.) Since there is a natural and not insignificant variability in one's scores, taking the test several times and submitting only the highest scores in each section also makes very good sense.

The big question, however, is how predictive of success in college one's SAT score is. The answer is a resounding somewhat.

There have been a variety of studies focusing on different groups of students and employing different protocols, controls, and assumptions. Most of them find that the correlation between SAT scores and college grades is not overwhelming. I surmise that the association appears rather weaker than it is because colleges usually accept students from a fairly narrow swath of the SAT spectrum. The SAT scores of students at Ivy League schools are considerably higher than those of community college students, yet both sets of students might have similar college grade distributions at their respective institutions. If *both* sets of students were admitted to Ivy League schools, however, I don't doubt there would be a considerably stronger correlation between SAT scores and college grades at these schools.

This is a general phenomenon: The degree of correlation between two variables depends critically on the range of the variables considered. Football players in the NFL are heavier, on average, than college football players. Weight is clearly significant in football, but one doesn't expect to see as close a correlation between weight and success within the NFL as there would be if college players and NFL players all played professional football.

And just as there are many dimensions of football ability that aren't measured by poundage, there are many dimensions of scholastic ability that aren't measured by the SAT. Concentrated work over an extended period is certainly one of them; the premium the SAT places on speed is especially difficult to defend. Anytime one tries to collapse a multifaceted, amorphous concept–physical beauty, political orientation, scholastic achievement, moral worth–along a linear scale, one is going to lose important information. I knew someone in graduate school in mathematics who scored around 400 in the math portion of the SAT (and comparably low on the Graduate Record Exam) and who now holds a prestigious endowed professorship in the subject. He is far from being alone, but if I knew only two things about a student–high school grades and SAT scores–I'd weigh the latter more heavily.

In the absence of a national curriculum or national standards, news coverage of the SAT helps provide a locus for discussion of educational policies and issues. Political reporters frequently build their stories around White House press releases, medical reporters wait for the latest edition of the *Journal of the American Medical Association*, and education reporters have the news on the latest SAT. Or as the statement might appear in the dreaded analogy segment of the test, political reporter : White House press release :: medical reporter : *JAMA* :: education reporter : SAT.

And those who think no discussion of the SAT is complete without a mathematics problem might consider the following, taken from the difficult portion of a recent test: In the correctly solved additions below, each of the five letters represents a different digit, EA being a two-digit number. What is the value of B + D if

$$
\begin{array}{cc}
  A & C \\
  + B & + D \\
  \hline
  C & E A ?^{*}
\end{array}
$$

---

*Combining the two additions yields A + B + C + D = C + EA. If we cancel the C's from both sides of this equation, we obtain A + B + D = EA, and thus B + D = EA − A. The two-digit number EA equals 10 × E + A, and so EA − A equals (10 × E + A) − A, or simply 10 × E. Since the digit E must be 1, (B + D) = 10 × 1, or just plain 10. (There are other approaches as well.)

Guns Will Soon Kill More
Than Cars

Comparability and Intensity

A recent spate of stories announces that guns will soon kill more people than do cars, the present number-one cause of injury-related deaths. The two graphs are projected to cross each other in the mid-1990s when, it's to be imagined, some safety-engineered car will function just long enough to participate in a drive-by shooting.

Although in favor of stricter gun control, I find these headlines a bit misleading. The Centers for Disease Control reports approximately 43,500 deaths in motor vehicle accidents and 38,300 deaths from firearms in 1991, the former number slowly decreasing, the latter increasing. But firearm deaths are almost always intentional. Only 4 percent of the 38,300 deaths from firearms were accidents, while 47 percent were homicides, 48 percent suicides, and the remaining 1 percent undetermined.* Related to this point of intentionality is the fact that opposition to automobile-safety standards is qualitatively unlike opposition to gun control.

The age and ethnic distribution of firearm victims also differs from that of car accident victims. Blacks and Hispanics between the ages of fifteen and thirty-four are, respectively, 4.7 and 1.9 times as likely to be killed by guns as are non-Hispanic whites. And for blacks in this age group, the firearm death rate is already 3.1 times the car accident death rate, 70 percent of the deaths being from homicides.

---

*If a randomly chosen person adds up the probabilities that each of the 5½ billion other people in the world will kill her, the sum, even in this violence-prone society, is still less than the probability that she'll kill herself.

However sliced, the figures on firearm deaths are more depressing and tragic than are those on car accidents. Why, then, are more stringent controls not placed on the purchasing of handguns and assault rifles? We place many constraints on driving and drivers, and polls have repeatedly shown that most people favor more controls on gunning and gunners as well. The question is complex, but a simple mathematical observation sheds some light.

Consider a scenario in which 20 percent of the electorate is opposed to stricter gun controls, three-fourths of them (15 percent of the electorate) so strongly that they will vote against any candidate who supports controls. Suppose further that the 80 percent of the electorate in favor of tighter controls do not feel as intensely. Let's say that only one-twentieth of them (4 percent of the electorate) will treat this as a litmus issue and vote against any candidate who opposes controls. Given these assumptions, it's not hard to imagine a prudent politician trying either to ignore the issue entirely or else to adopt a perfunctory opposition to gun control. Doing so loses him 4 percent of the vote; coming out for gun control loses him 15 percent of the vote. The difference of 11 percent would be significant in most elections. These single-issue voters—15 percent and 4 percent of the electorate, respectively—are often more determinative than are the 20 percent and 80 percent, respectively, from which they come.

Different issues arouse different degrees of intensity, however, and the issue of automobile safety is pivotal for an extremely small percentage of voters. More informative than a simple poll on an issue would be a weighted poll in which each vote is multiplied by a numerical measure of the voter's intensity (willingness to let the issue decide his or her vote). These weighted votes would then be summed to obtain an estimate of the issue's electoral potency. With deep commiseration for pollsters, I leave unsolved the more intractable problems of actually measuring voters' intensity and of determining how disparate issues are to be compared.

Note finally that the intensity of single-issue voters is a bit reminiscent of a possible prospect under Lani Guinier's cumulative-voting proposal. Instead of voters casting all their votes for a single candidate, they subordinate all their interests to a single overriding one.

Abortion Activists Bomb Clinic
Prohibitions and Arithmetical Arguments

Controversial issues like gun control often lead powerful organizations to take inflexible positions, making it very difficult for novel approaches and arguments to be heard. Since they want to be valued by the group, members freely express opinions in line with what they perceive to be the group's attitudes and tend to suppress those that run counter to those of the group. A prejudicial breeze soon develops and brings with it leaders who are more extreme than the average member. One simpleminded way to resist such constriction of options is to put forward as many positions and arguments as one can in the hope of freeing discussion and allowing a middle ground to form. Because they tend to be abstract, arithmetical arguments are well suited to this purpose.

Thus, when I read of implacable opponents of abortion who terrorize women and doctors at abortion clinics, for example, I sometimes try to imagine scenarios that would undermine the belief of *some* of them in the absolute inviolability of the fetus's right to life. Pro-life groups sometimes employ extreme arguments in their verbal skirmishes with pro-choice groups (and vice versa). If an abortion at three months is okay, why not one at six? And if one at six months is acceptable, why not kill infants, or the very old for that matter? Again, such arguments have their place if they induce a fresh rethinking of positions. A brief sketch of one from a pro-choice perspective follows, in the form of an imaginary news story:

Myrtle Jones, president of the pro-choice group Sense Not Sin, wondered at a rally yesterday what position abortion opponents might take if two facts about the world were to change. Ms. Jones asked her listeners to make two assumptions. The first is that, for some indeterminate reason—a virus, a hole in the ozone layer, some food additive or poison—women regularly become pregnant with thirty to fifty fetuses at a time. The second is that advances in neonatal technology make it possible to save easily some or all of these fetuses a few months after conception, but nonintervention at this time leads to the death of all the fetuses.

Opponents of abortion who believe that all fetuses have an absolute right to life would presumably opt for intervention, argued Ms. Jones. Their choice would thus be either to adhere to their position and be overwhelmed by a population explosion of unprecedented magnitude or else to act to save only one or a few of the fetuses. The latter choice would, Ms. Jones stressed, be tantamount to abortion, since all the fetuses are viable: "It would, nevertheless, take someone quite doctrinaire to opt to have the birthrate increase, at least initially, by a factor of thirty to fifty."

This is obviously not a knock-down airtight argument (although delivered to the right audience, it might result in knock-downs). Since it might be described as coming from a leftish direction (assuming the idea of a political spectrum has any validity), let me offer a more rightish example of a harmful narrowing of the terms of debate. The offender is a segment of the antismoking movement. More than 400,000 Americans die annually from the effects of smoking, but there is some intriguing evidence that the number could be drastically reduced by the widespread use of smokeless chewing tobacco. Professors Brad Radu and Philip Cole recently published a note in *Nature* in which they claimed that the average life expectancy for a thirty-five-year-old smokeless tobacco user would be fifteen days shorter than that for a thirty-five-year-old nonsmoker. This is in contrast to 7.8 years lost by smokers. The authors estimate that a wholesale switch to smokeless tobacco would result in a 98 percent reduction in tobacco-related deaths.

Since a small amount of tobacco lasts all day, tobacco companies would likely oppose smokeless chewing tobacco. There has already been strong opposition to it from *some* antismoking groups because of an increase in the risk of oral cancer (which is much rarer than lung cancer, emphysema, and heart disease). I suspect that another reason is a certain misguided sense of moral purity—not unlike opposing the use of condoms because, unlike abstinence, they're not 100 percent effective. If the numbers presented here are confirmed, however, recommending a switch to smokeless tobacco for those smokers (and only those) who can't quit would seem like sound public policy.

Including such bits of unconventional, arithmetically flavored reporting more often would make newspaper coverage of abortion, smoking, and other contentious issues less tiresome and might even kindle a more thoughtful response to them. There is almost always a greater variety of positions on any issue than ever make it into newsprint.

DNA Fingers Murderer
Life, Death, and Conditional Probability

The probability of someone speaking English given that (or on the condition that) he or she is an American citizen is, let's assume, 95 percent. But the probability that someone is an American given that he or she speaks English is much less—say, 20 percent. The conditional probability that someone is wealthy given that he or she is a cardiologist is very high. The converse conditional probability that someone is a cardiologist given that he or she is wealthy is very low.

An understanding of this mathematical notion of *conditional probability* is crucial to the proper interpreting of newspaper crime reports. In particular, it is at the base of what has been called the prosecutor's paradox, but should probably be termed the defense attorney's headache. Here's the story. A fingerprint or DNA fragment from the scene of a murder matches that of Mr. Smith. Newspaper headlines proclaim that the probability of a match with an innocent person is one in a million, let's say. Yet the more relevant conditional probability is the likelihood that a person is innocent given that his or her prints match the sample from the crime scene.

Let's get numerical. Imagine that this crime was committed in a city of approximately two million people. The year is 2001 and all the city's residents have records of their DNA or fingerprints on file. Assume further that three residents of the city have prints that closely match those at the murder scene; in practice, such prints are always a bit hazy and subject to interpretation. Two of these three people are innocent, the third guilty. Thus the conditional probability of a print match, given that a person is innocent, is two out of two million or, equivalently, one in a million. By contrast, the conditional probability that a person is innocent, given that his prints match those at the crime scene, is *two in three*; this latter probability constitutes more than reasonable doubt. Circumstantial evidence or motive should therefore always be sought to bolster forensic evidence.

Being clear about which conditional probability is the relevant one is crucial even in cases not involving high-tech detective work. The probability, for example, that an innocent person fits a complex verbal description of the perpetrator may be exceedingly small. But what, we should always ask ourselves, is the probability that someone fitting the description is innocent? That is what the jury must decide. Likewise, the probability that an honest poker player has been dealt a full house is about $1/7$th of 1 percent—much smaller, presumably, than the probability that someone who is dealt a full house is honest.

This little discussion elaborates a variant of one of the most basic distinctions in logic: "if A, then B" and "if B, then A." Relevant here is the work of the psychologist Peter Wason, who presented subjects with four cards having the symbols A, D, 3, and 7 on one side and told them that each card had a number on one side and a letter on the other. He then asked which of the four cards needed to be turned over in order to establish the following rule: *If a card has an A on one side, then it has a 3 on the other.* Most subjects picked the A and 3 cards. The correct answer is the A and 7 cards. Which cards would need to be turned over to establish this rule: *If a card has a 3 on one side, then it has an A on the other?*

---

*The 3 and the D.

Darts Trounce the Pros

Luck and the Market

The sports pages of the January 11, 1994, *Wall Street Journal* reported that the Darts scored a smashing victory over the Pros in their ongoing series. The sport, of course, is stock picking, and the darts are just that—random selections by dart—while the pros are a rotating collection of market experts. The Darts averaged a 42 percent gain in the period July 7, 1993, to December 31, 1993, compared with 8 percent for the Dow Jones and 2.2 percent for the experts. As of the date of the article (the pattern has continued since then), the Pros had twenty-five wins to the Darts' eighteen in a series of overlapping six-month contests. Their lead over the Dow Jones was a narrow 22 to 21.

Mutual funds also display this chancy behavior. Although not as volatile and unpredictable as individual stocks, they're still quite likely to turn up in the top quarter one year and in the bottom quarter the next (as a study of *Barron's* mutual fund watch will bear out).* Since the market is probably not perfectly efficient, information is not immediately reflected in prices, and none of us is immune to greed and fear, the pros fare a bit better overall, especially if we look at batting averages (percentage gains) rather than games won. Nevertheless, the differences are not sufficient as to be inconsistent with the thesis that stocks behave in an essentially random way.

---

*A similar point can be made about many rankings. If one examines U.S. airlines' safety records over rolling three-year intervals, one finds that it's common for the "safest" airline during one interval to be the "most dangerous" during the next.

Whatever one's position, there are enough similarities between stocks' performances and completely random processes for investors to benefit from a look at the latter. Humor this mathematician then, if you will, and imagine flipping a penny one thousand times in succession and obtaining some sequence of heads and tails. (Incidentally, make sure you imagine flipping it and not spinning it. *Spinning* a penny results in heads only about 30 percent of the time, not the 50 percent that results from flipping.) If the coin and flip are fair, or even if the flip is biased to reflect the general upward drift of the market, there are a number of curious facts about such random sequences that could save you brokers' commissions. For one, if you were to keep track of the proportion of the time that the number of heads exceeded the number of tails, you might be surprised to discover that it is rarely close to $1/2$.

Picture two players, Dim and Dum (the offspring of father Dow), who flip a penny once a day and who bet on heads and tails, respectively. Dim is ahead at any given time if there have been more heads up until that time, while Dum is ahead if there have been more tails. Although Dim and Dum are each equally likely to be in the lead at any given time, it can be proved that one of them will probably be ahead almost the whole time. Thus, if there have been one thousand coin flips, the chances are considerably greater that Dim (or Dum) has been ahead more than 90 percent of the time than that he's been ahead between 45 percent and 55 percent of the time! Likewise, it's considerably more likely that Dim (or Dum) has been in the lead more than 98 percent of the time, say, than that he's been ahead between 49 and 51 percent of the time.

This result is counterintuitive because we tend to think of deviations from the mean as being somehow bound by a rubber band: the greater the deviation, the greater the restoring force toward the mean. But even if the penny landed heads 525 times and tails 475 times, for example, the difference between its heads total and its tails total is just as likely to grow as to shrink with further flips. This is true despite the fact that the proportion of heads does approach $1/2$ as the number of flips increases. (The gambler's fallacy and this more subtle consequence of it should be distinguished from another phenomenon, regression to the mean, which is valid. If the penny were flipped one thousand more times, it is more probable than not that the number of heads on the second thousand flips would be smaller than 525.)

If even fair coins behave so oddly, one should expect that some stock pickers will come to be known as losers and others as winners without there being any real difference between them other than luck. If Dim and Dum have won, respectively, 525 and 475 trials, Dim will likely be profiled in the business pages as a visionary and Dum disdained as a plodder. Winners (and losers) are often just people who get stuck on the right (or wrong) side of even. It can sometimes take a very long time for the lead to switch hands.

Another counterintuitive aspect of coin flipping concerns the surprising number of consecutive runs of heads or tails of various lengths. If Dim and Dum continue to flip a penny every day to determine who pays for their *Wall Street Journals*, then it's more likely than not that at some time within about nine weeks, Dim will have won five papers in a row, as will have Dum. And at some period within about five to six years, it's likely that each will have won ten papers in a row (assuming the paper comes out seven days a week).

Random events can frequently seem quite ordered. The following diagram is a computer printout of a random sequence of 260 H's and T's (52 groups of 5), each letter appearing with the probability $1/2$. Note the number of runs and the way there seem to be clusters and other patterns. If you feel compelled to account for these, you will have to invent explanations that will of necessity be quite false.

THHHT	HHHTH	HHHTT	HHTHH	THTTH	THTTT	THTHH
TTTHH	HTHTH	HHHHH	HHHTH	HHTHT	HHHHT	HTTHH
HTTTH	HHHHT	THHTT	THHTT	TTTHH	TTHHH	HHHTH
HHHTT	HHHHT	THHTH	HTTHH	THTHT	THHHT	HHTHH
HHTHH	HTHHT	HHHHH	HHHHT	HHHHH	TTTTT	HTTHH
HTTHH	HHTTH	TTHTH	HHTHH	HHTTT	THTHT	HHTHH
HTTHH	TTTTH	HHHHT	TTTHH	HHTHH	TTHHH	HHHTH
HTTTT	TTटHT	HHHHH				

Another easy exercise illustrates how patterns seem to emerge from pure randomness. Take a piece of white paper and partition it into squares so that it looks like a checkerboard. Flip a coin and color the upper-left corner red if it lands heads and blue if it lands tails. Proceed to the next square and repeat. When you've colored the whole paper, look it over for patterns and connected clumps of similarly colored squares. What do you think would be the effect of doing this on a three-dimensional checkerboard?*

Now, with these random clumping patterns in mind, think of the standard pronouncements of newspaper stock analysts. The daily ups and downs of a particular stock or of the stock market in general may not be as thoroughly random as these H's and T's are, but it's safe to say that there is an extremely large element of chance involved. (Increasing the probability of H to slightly more than $1/2$ to simulate roughly the positive trend of the market doesn't alter any of this.) One never hears of chance, however, in the neat post hoc analyses that follow each market's close. Chartists and technicians invariably discern heads and shoulders, triple tops and bottoms, wedges and channels in the behavior of individual stocks. Fundamentals analysts point to a company's dividends, profits/earnings ratios, and other presumed determinants of its stocks. Market commentators always have a familiar cast of characters to which they can point to explain any rally or decline.

There's invariably profit taking or the federal deficit or something or other to account for a bearish turn, and improved corporate earnings or interest rates or whatever to account for a bullish one. To explain more persistent trends, it's always tempting to point to unsettled politics. I'm writing at the end of March 1994, and the stock market has fallen 350 points in the last couple of months. The economy seems to be in better shape than it's been in for a long time, yet instead of displaying a modicum of uncertainty, columnists have confidently attributed its problems variously to North Korea's having the bomb, Yeltsin's drinking too much vodka, Alan Greenspan's

---

*In the case of the plane checkerboard, note that most squares share a line with four other squares and at least a point with eight others. The connected clumps of like-colored cubes would be larger and more extensive on a three-dimensional checkerboard since here most cubes would share a face with six other cubes and at least a point of contact with twenty-six other cubes. In higher dimensions, almost everything would be connected via some route to almost everything else. Ponder the implications if you will.

being a nasty man, and Hillary Clinton's liking pork bellies.

Almost never does a stock pundit say that the market's or a particular stock's activity for the day or the week or the month was largely a result of random fluctuations. One will never declare something like, "I'm damned if I know what caused this." (I'll gloss over the possibility of subtle dynamical patterns being eventually unearthed by chaoticians.)

The business pages, companies' annual reports, sales records, and other widely available statistics provide such a wealth of data from which to fashion sales pitches that it's not difficult for a stock picker to put on a good face. As with coin flips, he or she can readily focus on a run of good luck or a comparison with a rival that indicates marginally superior performance over a given period of time. All that's necessary is a little filtering of the sea of numbers that washes over us; it's easier than flipping a coin!

Cellular Phones Tied to
Brain Cancer

Multiplication, Health, and Business

It is by turns amusing and depressing to track the way descriptions of numerical relations depend on their authors' intentions. To make a quantity appear large, for example, a consumer group, political group, or business advertiser might stress a linear measure of its size. To make it appear small, it might stress its volume. Thus, although a single tower of nickels stretching from sea level to the height of Mount Everest would contain more than four million coins, you can easily verify that this pile would fit comfortably into a cubical box about 6 feet to a side. And spacious cubical apartments (20 feet on a side) for every human being on Earth could fit comfortably into the Grand Canyon. By contrast, if all living humans were placed end to end, they would extend to the moon and back more than eight times.

A related equivocation arises when one is discussing diseases, accidents, or other misfortunes and their consequences. If one wishes to emphasize the severity of a problem, one will usually talk about the number of people afflicted nationally. If one wants to downplay the problem, one will probably speak about the incidence rate. Hence, if 1 out of 100,000 people suffers from some malady, there will be 2,500 cases nationwide. The latter figure seems more alarming and will be stressed by maximizers. Dramatizing the situations of a few of these 2,500 people by publishing or televising interviews of their families and friends will further underscore the problem. Minimizers, on the other hand, might invoke the image of a crowded baseball stadium during a World Series game and then point out that only one person in *two* such stadiums suffers from the misfortune in question.

A paradigm for many recent health and business scares, the case of cellular telephones and brain cancer is illustrative. Concerns were raised a few years ago when a guest on a national talk show blamed his wife's recent death from brain cancer on her use of such phones. The appearance, the allegation of a causal connection, the concomitant lawsuit, and the ensuing media delirium led to fear, confusion, and a decline in the stock prices of the companies that manufacture cellular phones. Anxiety, dread, and the power of dramatic anecdote clouded the mathematical distinction between the rate of incidence and the absolute number of instances.

In one peculiar view, the "data" suggested a specious mathematical argument that seems to demonstrate that these devices actually *inhibit* the formation of brain tumors. The argument, direct from the Freedonia School of Media, Law, and Muddled Policy: There are an estimated 10 million users of cellular phones in this country, and the incidence rate for brain cancer among all Americans is 6 per 100,000 annually. Multiplying 10 million by $6/100,000$, we determine that approximately 600 cases of brain tumor should be expected annually among users of cellular phones. Since the evidence for an association between cancer and cellular phones consists of only a handful of people, far short of even one year's contribution of 600, we conclude that cellular phones effectively ward off brain tumors. Absurd to be sure, but no more so (in fact, less so) than the original hysteria.

Hysteria may have also been a factor in some of the stories on the harmful effects of breast implants. The most recent and careful study, undertaken by the Mayo Clinic and reported in the *New England Journal of Medicine*, indicates that, despite a four-billion-dollar judgment against the manufacturers for failure to test them adequately, the implants do not cause the host of connective tissue disorders attributed to them. Although the 2,250 women in this study—one-third with implants, two-thirds without—is not a large enough number to be definitive, seldom has the disjunction between science and law seemed so stark.

It's easier and more natural to react emotionally than it is to deal dispassionately with statistics or, for that matter, with fractions, percentages, and decimals. The media (actually, all of us) frequently solve this problem by leaving numbers out of stories and hiding behind such evasive words as "many" or "uncommon," which are almost completely devoid of meaning. And when statistics are included in an article, the intense coverage afforded certain topics sometimes undermines their import (or lack thereof). In the journal *Chance*, the statistician Arnold Barnett examined how frequently different modes of death became page-one stories in the *New York Times* during a two-year period. He found that they ranged from .02 stories per 1,000 annual deaths from cancer to 138 stories per 1,000 annual deaths from air crashes.

Even in business stories, where there are fewer psychological obstacles to a more quantitative approach, it's sometimes difficult to determine whether a cost rise of 5 percent, for example, is based on revenues, last year's costs, or what. More subtle complications also arise with some regularity. Imagine buying 100 pounds of potatoes, for example, and being told that they're 99 percent water. After the potatoes have been left outdoors for a day or two, you're told that they're now 98 percent water. At first, it doesn't seem difficult to determine the weight of the slightly dehydrated potatoes, but I've discovered that few people can do so.*

You're probably not particularly interested in potatoes, so let's say that your stockbroker relieves you of a large sum of money and informs you that only 1 percent of your investment goes to him. Later the value of your investment declines, but your stockbroker, knowing all about potatoes, assures you that his share still constitutes only 2 percent of the present value of your investment. Losing half your money is not necessarily a matter of small potatoes.

Ideally, readers and reporters would have the rudimentary numeracy to be able to translate from rates to absolute numbers and back, or to convert from one unit to another (from length to volume,

---

*Not worrying too much about the structure of potatoes, note that the 100 pounds was 99 pounds of water and 1 pound of potato essence. It now weighs X pounds and is 98 percent water and 2 percent potato essence. Thus 2 percent of X is 1 pound. Since $.02X = 1$, $X = 50$ pounds. The answer is that the potatoes now weigh just 50 pounds.

for example), or to use percentages with accuracy and clarity (two desiderata often in conflict). They would also have the basic factual knowledge and the appropriate definitions essential to making reasonable estimates: the population of the metropolitan region, the United States, and the world; certain geographical distances and socioeconomic measures; some feel for common magnitudes; business figures of significance; and so on. With this facility, reporters could somewhere in their articles provide various equivalent formulations of the numbers under discussion, or readers could determine these for themselves. As it is, the formulations provided by news sources who often have their own axes to grind are generally the only ones presented in a story.

Consider the number of abused women, for example. The figures cited in newspapers vary dramatically depending upon what classificational criteria are used. Is the number based on police crime reports, sampling interviews, or the general impressions of the ideologically engaged? And how much credence should we give to reports of immigrants who either receive $44 billion more in government benefits than they pay in taxes or else pay $25 billion more in taxes than they receive in benefits, depending on what you read? Or consider that the estimated percentage of men who are homosexual ranges from the recent Guttmacher Institute's report of 1 percent to the Center for Health Policy's claim of 20 percent (which counts anyone who has ever felt any homosexual attraction since the age of fifteen). Similar caveats apply to the number of homeless people in America, which ranges from 200,000 to 7,000,000. Either figure or any intermediate one might conceivably appear in a story or headline; much less likely to appear will be an account of the operational definitions used or of other numbers that would follow from the ones reported.

In an insightful essay, the *Newsweek* columnist Jerry Adler wrote that journalists "consider their job done when they find a credible source." Unfortunately, it's not. Benchmark figures, operational definitions, and simple arithmetic should be a part of every major story or should at least appear frequently in the ongoing coverage. Without them, we all tend to be unduly swayed by the dramatic, the graphic, the visceral.

GM Trucks Explode on Side Collision

From Pity to Policy

There isn't, I'm sure, but there sometimes appears to be an informal alliance among journalists, lawyers, and people who claim to have been injured by faulty products or services. (Incidentally, there are now 864,000 lawyers in the United States, according to the American Bar Association. This is an increase of 59 percent since 1980 and 143 percent since 1971.) Cellular phones, hotel safety, pickup trucks, childhood vaccines, campus security. It's almost impossible to pick up a newspaper or watch television without being confronted with the doleful case of some recently bereaved family.

Some of the defendants in such cases are reprehensible and undoubtedly guilty, but this trend seems unfair to them as a group. No one can argue effectively with the raw anguish of families and friends, but I would like to see, just once, some earnest news-magazine anchor say, "This is all very, very tragic, but what policy would you alter that would make tragedies of this sort less likely and would not simultaneously increase the likelihood of other, less visible types of tragedy?" (On further thought, I really wouldn't want to see this.) Instead, the grief of the victims and the pity it engenders are too often used as a rationale for the *impossible* demand that there never be any risk or as a shield against criticism that the news coverage is tendentious.

These endlessly recurring stories, whether they pop up in the business pages, in the local news sections, on daytime talk shows, or on the burgeoning number of electronic newsmagazines on TV, are frequently based on the testimony of a handful of people, few of whom are disinterested parties. Scientists are seldom heard from, lawyers almost always. Uncertainty is a common and sometimes unavoidable state of affairs in science, but judges, lawyers, and juries often act as if every question has a definitive answer if only the witness thinks hard enough, the experts calculate long enough, and cover-ups are exposed.

And unlike the plaintiffs, the defendants find it advisable for legal reasons not to discuss the case, thus leaving the impression in the reading or viewing public that they're guilty or at least heartless. If they or neutral parties are heard from, their comments are usually sandwiched between testimony from the victims or their lawyers, giving the plaintiff both the first and last word. Moreover, if the defendants' side of the story depends in any crucial way on something technical, one can be sure it will be too boring to make compelling reading or viewing. Pathos makes for both a better visual and a better read.

I have no knowledge, for example, of the safety considerations that General Motors engineers employed when designing their trucks. It is conceivable, however, that they concluded that locating the fuel tanks outside the frame brought about a minuscule increase in the probability of a fire on impact, while significantly enhancing safety in other ways. They might have concluded that the trade-off was reasonable.* To try to give this abstract point visual impact comparable to that of a devastated family would be futile. One might start by showing some of the thousands of happy faces who, often unknowingly, avoided death or injury because of this possible design decision.

There is another psychological foible that a plaintiff's lawyers can exploit. In assigning causes, people are much more likely to attribute an event that has momentous or emotional implications to an agent rather than to chance. In one experiment, for example, a

---

*The number of deaths resulting from the fuel tank placement is orders of magnitude less than the 4,000 annual deaths attributable to an absence of passenger-seat airbags.

group of subjects is told that a man parked his car on an incline, after which it rolled down into a fire hydrant. Another group is told that the car rolled into a pedestrian. The members of the first group generally view the event as an accident; the second group holds the driver responsible. It's a charming superstition that significant consequences must be the result of significant negligence.

Providing knowledge of the potential problems and possible criminality associated with some service or manufactured item is in the public interest. Reporters who write and papers that publish such exposés are to be applauded; there should be more of them. What, however, is the rationale for showing so much pain so regularly? Is anyone surprised that people cry when their loved ones suffer or die? (Why not take things a step further and have a show devoted exclusively to the keening and wailing at funerals? *Unsolved Miseries* might be a winner.) The emotionally evocative, dramatic, and concrete presentation of a victim's plight can seal the public relations case, if not the legal one when defendants settle out of court to end the publicity. That's a cost to all of us. So is the cynicism that saps our outrage at the most egregious bad guys. And so, finally, is the numbing associated with our easy voyeurism.

The $32 Billion Pepsi Challenge
Advertising and Numerical Craftiness

The lifeblood of the newspaper business, advertising is sometimes artistic, often instructive, and always seductive. Since its appeal is usually to our emotions and because most people know how the game is played, it is perhaps a little disingenuous to nitpick about this or that bit of misleading copy. Still, my professorial compulsiveness requires that I touch on the topic.

As in magic, the problem with ads isn't so much what is said or demonstrated as what we infer from what is said or demonstrated,* and the best way to induce false inferences is to sketch alluring pictures and leave out crucial bits of information. You can have this appliance for as little as $125 a month. But how many months? What's the down payment? And are there balloon payments or other charges? A survey shows that this medication works more quickly. Than what does it work more quickly? Why is quick action important? Is it quick, but relatively ineffective? This nutrient is essential to good health. Are we suffering from a lack of it? Are there other sources of it? Can we have too much of it?

---

*Speaking of mathemagic, recall the story about the three men who check into a hotel and rent a room for $60. After they go to their room, the manager realizes that the room costs only $55 and that he's overcharged them. He gives $5 to the bellhop and directs him to give it back to the three men. Not knowing how to divide the $5 evenly, the bellhop decides to give $1 to each of the three men and pockets the remaining $2 for himself. Later that night the bellhop realizes that the men each paid $19 ($20 minus the $1 they received from him). Thus, since the $57 the men paid plus the $2 that he took for himself makes $59, the bellhop wonders what happened to the missing dollar.

These omissions combined with standard tricks involving graphs (the most minuscule variation can be made to appear significant with an appropriate choice of scale), hocus-pocus with percentages and the like (this cream reduces pimples by 200 percent, suggesting, it would seem, that little indentations will appear where there used to be bumps), and the inclusion of an ample supply of irrelevancies (beautiful models, evocative situations, majestic settings) make reading ads almost fun. Every day there is a new game of Spot the Sleight.

This held true for television as well until the invention of the muting button on the remote control, which, interestingly, has flourished without itself being advertised very much. In every media there are the same sorts of misdirection, perennials such as the celebrity endorsement, free samples, two-for-one sales, and that old baffler, the $999 item, not its $1,000 cousin. An interesting twist on an old trick is the ad that proclaims that you'll pay 50 percent more at a competitor's store rather than announce an equivalent, but less impressive, 33-percent-off sale.

More inventive subterfuges surface periodically, or maybe I just imagine them. An otherwise unobjectionable ad for a product stated that a laboratory test showed no statistically significant difference between the product and the higher-priced market leader. "Statistically significant" was in boldface type. I don't know anything about the product in question, but it struck me that such a claim could easily be true, but almost empty. To make it valid, the company would merely have to commission a test with a very small sample. From such a sample, it would be impossible to draw *any* conclusions that are statistically significant. Only the grossest differences between the two products would be discernible.

Or consider the ad for a new restaurant that offers an all-you-can-eat special for weekend brunches. Noting the sheer volume of food some people can consume, I've often thought that if such an establishment offered an all-you-*may*-eat brunch, it could legitimately send a bouncer over to gluttonous parties to inform them that they have eaten all they may eat.

A less fanciful example concerns claims made by unctuous TV

pitchmen (and appearing in print as well) that hospital bills can be $50,000 or more but that Medicare pays only 55 percent of bills on average. A possible racket springs to mind. For illustration's sake, assume that the average deductible for Medicare is $500. Assume further that though some medical bills are $100,000 or more, most are only a few hundred dollars. Now divide the claims into two sets, those in which the bill is, let's say, $2,000 or under and those in which it exceeds $2,000.

For each bill in the first group, find the percentage of the bill that Medicare pays. Since the deductible is $500 and bills in this category are no more than $2,000 and usually much smaller, the average percentage paid by Medicare is, let's suppose, only 20 percent. For each bill in the other category, do the same thing. Since some of these bills are huge and all are over $2,000, the average percentage paid by Medicare is here much higher—say, 90 percent. Averaging these two averages would result in the phony average of 55 percent and lead to the unfounded claim.

Whether deceptive or straightforward, advertising campaigns often fail too, but seldom is this because companies slip up and make a mathematical claim that hurts them. One real estate development company advertised that an investment with it would grow logarithmically (that is, very, very slowly—surely not what it meant to convey) over the years, but I suspect that its failure was independent of this mathematical gaffe. And a car ad that bragged of the highest price/performance ratio in the industry probably wouldn't even hurt sales.

One promotion that constitutes a sort of exception is Pepsi's catastrophic lottery in the Philippines a few years ago. On May 25, 1992, the nightly news in Manila announced the winner of its well-publicized bottle-cap contest. Anyone holding the winning cap number, 349, would receive one million pesos, about $40,000. There is nothing unusual about promotions like this one, and, at least in the short run, they do seem to increase business. The odd thing about this particular case, as the *Los Angeles Times* reported, is that through some sort of mix-up—computer or otherwise—Pepsi had put out

800,000 bottle caps with the number 349! Hundreds of thousands of people, rich and poor alike, demanded their money–potentially as much as $32 billion. After some violence and an offer from Pepsi of $20 per winning number, the situation seems to have faded away.

Pointing out the arithmetical abuses in advertising is a little like being a garbage collector: You've got to do it regularly or the stuff piles up. Unlike garbage, however, most of it smells good, and a rea-sonable percentage of it even is. In any case, however it's character-ized, advertising is what makes the newspaper business a lucrative one.

Brief Fads Dominate Toy Industry

S-Curves and Novelty

The toy business has interested me ever since, late in the Rubik's Cube craze of the 1980s, I spent a lot of time and money patenting a variant of the cube that I called About Face. On the six sides of the cube are human faces that remain faces when viewed upside down. About Face was both easier and harder to master than the standard cube. Because the ears, chins, foreheads, and other features could be manipulated independently, it was quite easy to form a variety of faces, some of them famous. However, because the orientation of the center square made a difference (otherwise the eyes might be vertical), it was more difficult to reattain the original configuration than it was in the Rubik's Cube. Needless to say, the fad died and with it the prospects for About Face.

There is, in any case, a mathematical curve that is brought to mind by the headline that titles this segment and by fads in general. I'm not referring to the normal curve in statistics, the curve for the exponential growth of money, or the parabolic trajectory of a basketball, but rather to the S-shaped curve. This curve characterizes, or at least seems to characterize, a variety of phenomena, including the demand for new toys. Its shape can most easily be explained by imagining a few bacteria in a petri dish (see diagram). At first, the number of bacteria will increase at a rapid exponential rate because of the rich nutrient broth and the ample space in which to expand. Gradually, however, as the bacteria crowd each other, their rate of increase slows and the number of bacteria stabilizes.

The S-Curve

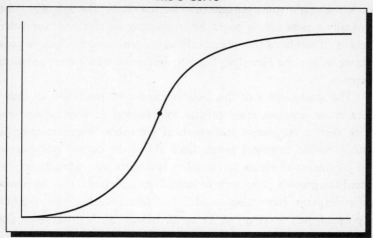

The S-curve describes many phenomena.
Growth begins to slow at point indicated.

Interestingly, this curve (sometimes called the logistic curve) appears to describe the growth of entities as disparate as Mozart's symphony production, the rise of airline traffic, new mainframe computer installations, and the building of Gothic cathedrals. If you can't think of more, the writer Cesare Marchetti and others have amassed a large number of them and speculated that there is a kind of universal principle governing many natural and human phenomena. What is especially provocative about the supposition is that in many of their examples, there doesn't seem to be anything analogous to the nutrients in the petri dish—no resources whose depletion leads to an end to exponential growth and a gradual leveling off.

But there is, I suggest, something that is being continuously depleted with time, and that is the (admittedly vague) sense of novelty. Our natural proclivity to focus on the unusual, the dramatic, and the new is strengthened immeasurably by newspapers and other media, but our interest wanes quickly as well. We're so fascinated by the sudden rise of the new celebrity, the spread of the titillating rumor, and the increasingly frequent accounts of some exotic condition or lurid crime that we forget the trivial fact that many phenomena have a limited life span. And given this life span, it shouldn't be too surprising if some phenomenon begins small, takes off, and then gradually tapers off. It might be interesting to plot the *cumulative* number of mentions that previously unknown persons, ideas, or fads receive in, say, the *New York Times* to determine which ones generate S-curves.

The mathematics of the S-curve cannot be predictive of these phenomena without more precise information. It may be nothing more than a suggestive mathematical metaphor. Supplementing it with plausible empirical assumptions about the curve's parameters and arguments about its applicability, biologists have accurately predicted the growth of bacteria in limited environments. Human population experts have also used it to forecast that the world's population will level off at 11 to 12 billion. Such predictions are probably not any more valid than those of the toy-marketing experts I consulted, but whatever its predictive power, the S-curve is also helpful in clarifying a point about indices.

We've all heard the television anchor with the authoritative voice intoning that such and such an index is declining (or rising) or seen the generic headline proclaiming THINGS GETTING WORSE (or BETTER). When are such pronouncements warranted? At what point can we say, for example, that we are witnessing a deterioration? Is it when the index is falling or when its ascent is slowing or its descent quickening? Arguments can be adduced for the last two positions. (For those who know calculus, the question is whether the first derivative is negative–the index is falling–or whether the second derivative is negative–either the ascent is slowing or the descent is quickening.) The point is that even in this very simple case, a mere decline in the relevant index needn't be a cause for despair nor need a rise be a cause for joy.

The point on the graph of the S-curve where it switches from being concave up (smiling) to concave down (frowning) is a critical one. This is where the growth, though still positive, begins to slow. If the quantity indexed is something desirable, then, in a sense, things start to get worse at this point. In another sense, they're just getting better more slowly. We must examine what the specific index in question is measuring to evaluate the situation.

One last point about these indices: Situations in which the value of such an index is rising most rapidly are often only superficially worrisome (or hopeful, depending upon what is being measured). An example is the increase in the incidence of AIDS in rural women over sixty years of age. If the baseline incidence is very low, a few new cases can result in stories announcing a dramatic rise in the relevant index.

Rates of change, rates of rates of change, and the relations among them constitute the bulk of the mathematical discipline of differential equations. It's noteworthy that the rudiments of the subject are implicit in seemingly straightforward news stories. Continuing on this topic, however, will lead us beyond the point where the S-curve of interest in it begins to level off.

Area Residents Respond to Story

Repetition, Repetition, Repetition

The man-on-the-street-reaction story provides a manifestation, alternately annoying and humorous, of the widespread tendency to present blather and call it news. The television version usually begins, "And now for a local response to these developments, we go to ... " Then "we" briefly visit a tavern, a school, a bus stop, and a department store, where various ordinary people (whatever that means) are confronted by a reporter and asked their opinion on some breaking story that has just been described to them. The situation always reminds me of people who, a day after an argument with their husband or wife, inform the spouse that everybody they've spoken to about the argument agrees with them. (The spouse's response should be, "If I had heard only what you described to them, I would have agreed with you, too.") Unless the proverbial man on the street has an independent view, a contrary perspective, or additional information, I don't care what he or she thinks.

Apropos is the Austrian philosopher Wittgenstein's account of a man who, being unsure of a story he reads in the newspaper, buys dozens of copies of the paper to corroborate the story. As the French philosopher Henri Bergson would have recognized, the repetitive way in which local-reaction stories (among other types) are constructed often makes them funny, instances of "the mechanical encrusted upon the living"–his characterization of humor. Even tragic events repeatedly enacted quickly begin to seem farcical.

Either a greater effort should be made to locate and interview local people with a new slant on the story or else the preface to these reaction pieces should be, "Here are some abbreviated, quite minor variations on our story recounted by an indiscriminate collection of neighborhood residents."

Researchers Look to Local News for Trends

The Present, the Future, and Ponzi Schemes

Even the most superficial reading of a newspaper reveals an important aspect of human psychology: our preoccupation with the short term. Essential to our survival, our myopic focus on the day's happenings can nevertheless cause problems for us. Evolution's favoring of organisms that respond to local or near-term events results in a steep temporal and spatial discount rate for distant or future events. The latter are discounted in the same way that money is. Suffering ordained for twenty years from now is, like a million-dollar debt due in twenty years, considerably easier to bear than is suffering scheduled for tomorrow.

The relevance of this to AIDS and other sexually transmitted diseases, gastronomic or other indulgences, financial excesses, radioactive wastes, and greenhouse effects is obvious. Environmental despoliation may even be conceived as a kind of global Ponzi scheme,* the early "investors" doing well, the later ones losing everything. Nevertheless, the "right" discount rates are not easy to determine and vary from case to case. It's not that we shouldn't prize the here and now, but, as residents of a global village whose actions can reverberate for a long time, we may need a Global Reserve Board to help decide on more rational discount rates to keep us from being our own Ponzis.

---

*In a classic Ponzi operation, the few early investors are paid off with the contributions of many later investors, who, in turn, are paid off by the contributions of still later ones, until the scheme collapses.

A quasi-mathematical illustration of the psychology involved is provided by Robert Louis Stevenson's "The Imp in the Bottle," the story of a genie in a bottle who will satisfy your every wish for love, money, and power. You can buy this amazing bottle for any amount that you care to offer. The only constraint is that when you are finished with the bottle, you must sell it for less than what you paid for it. If you don't sell it to someone for a lower price, you will lose everything and suffer everlasting torment in hell. What would you pay for such a bottle?

Certainly you won't pay 1 cent for it because then you won't be able to sell it for a lower price. You won't pay 2 cents for it either, since no one will buy it from you for 1 cent for the same reason. Neither will you pay 3 cents for it; the person to whom you would have to sell it for 2 cents wouldn't be able to sell it for 1 cent. A similar argument applies to a price of 4 cents, 5 cents, 6 cents, and so on. Mathematical induction can be used to formalize this argument, which proves conclusively that you shouldn't buy this magic bottle for any amount of money. Yet you would almost certainly buy it for $1,000. I know I would. At what point does the argument against buying the bottle become practically convincing?

The consequences of our decisions need not occur in the future for us to discount them. They can occur far away or after so many steps as to seem distant. Derek Parfit, in his book *Persons and Reasons*, discusses the case of someone strapped to a bed, with electrodes firmly attached to his temples. Rotation of a dial in another location imperceptibly increases the current in the electrodes. A free hamburger is offered as an incentive to anyone who twists the dial. Assuming it takes ten thousand people each rotating the dial once to electrocute the victim, what degree of guilt attaches to each individual dial twister? Do these tiny guilt bits accumulate in any moral bank account?

And what about the apocryphal story of the clever programmer at the Social Security Administration? The checks of senior citizens, the story goes, were calculated to a tenth of a cent and then rounded up or down to the nearest penny. The programmer skimmed this fraction of a cent off everyone's check and electronically directed it

to his own private account. Is this a victimless crime? Do these fractions of a penny add up to grand larceny?

Another relevant metaphor from mathematics is brought to
mind by this verb *add up*. We often want to figure the net effect of
decisions or actions made over time, distance, or some other dimension. The term suggests the mathematical operation of integration
whereby the values of quantities that vary over some dimension are
added up or summed. What, for example, is the future value of an
annuity or the cumulative impact of thousands of small environmental or personal abuses over time? In *War and Peace,* Tolstoy uses the
metaphor in discussing the integration of the private intentions of
thousands of soldiers into one large historical force.

Such integrations are, needless to say, incomparably more difficult than the integration of mathematical functions in calculus. Some
people try, however. By reading hundreds of newspapers from all
over the world, futurists such as John Naisbitt and Alvin Toffler
attempt to "add up" the causes and effects of countless local stories in
order to identify and project trends. Their hope is that in the process
they will limn a rough outline of tomorrow. The projections are usually along a straight line, interactions among the various trends are
commonly ignored, and unexpected developments, by definition, are
not taken into account. As with weather forecasters, the farther
ahead they predict, the less perspicacious they become.

Section 3

LIFESTYLE, SPIN, AND SOFT NEWS

The other one, the one called Borges, is the one things happen to.
—JORGE LUIS BORGES

Concern with self, reference to self, and promotion of self are much more common in popular culture and the media than ever before. And not just on daytime TV. According to a 1991 Gallup Poll, most of the 70 percent of Americans who look through a newspaper on any given day read the soft lifestyle, gossip, and daily magazine sections of the paper. Furthermore, this preoccupation with self transcends the lifestyle sections; spin control, public relations, and a concern with the media and celebrity play a role in every domain from business to international affairs. Not even natural disaster stories are immune; during the recent Los Angeles earthquakes there were headlines like QUAKE DOESN'T SPARE STARS' HOMES. Wouldn't the real story have been an earthquake that did skip over the houses of the rich and famous?

And newspaper op-eds, by their very nature, necessitate some reference to self. Exactly how much was answered by *The Nation*, which ran a story in January 1994 that ranked twenty-two *New York Times* and *Washington Post* op-ed and political columnists according to their use of the words *I, me,* and *myself.* Richard Cohen led the pack, averaging 13.4 self-mentions per column; he was followed by Meg Greenfield, with 8.4. Ellen Goodman, Anna Quindlen, David Broder, and William Safire weighed in at, respectively, 4.1, 3.7, 1.7, and 1.6. Mary McGrory scored only a .8, and George Will was self-effacement champ with .2. Can one brag about this distinction? Intrigued by these rankings, I had my word processor count the num-

ber of times I had used those self-referential words in various segments of this book; the result was approximately 1.6 times per 700
words, the length of the average column. Of course, there are problems with such a simple criterion, but I'll leave those to the reader.

Looked at from a mathematical perspective, this concentration
on self and the complexity it engenders bring to mind a number of
notions in logic and computer science. The issues discussed in this
section include celebrity profiles, presidential primary handicapping,
societal obsessions, spin-doctoring and the reporter's role in stories,
compressibility of news accounts, and the paradoxes and ironies that
lie just below the surface of some articles. (Recall the pollster who
asked, "To what do you attribute the ignorance and apathy of the
American voting public?" and was answered, "Don't know and don't
care.") I also discuss the notion of a person's or a group's "complexity
horizon" and the many aspects of modern life that are so intricately
tangled and convolutedly complicated as to be beyond that horizon.
Finally and perhaps fittingly, I include in this section on self another
personal interlude in which I average more I's and me's than Richard
Cohen.

A Cyberpunk Woody Allen

How to Write a Profile of the Fledgling Celebrity

It sounds like the punchline of a joke. The mathematician is the one who, when confronted with the role of money, sex, and power in the creation of celebrity, focuses instead on numbers, graphs, and logic. Suffering from this occupational myopia and possessing the tiniest smidgen of fame myself have prompted me to write this segment on the mathematical aspects of celebrity construction, focusing on the profile of the fledgling celebrity that often appears in the style sections of newspapers. So sit back and assume that for the duration you are a mathematically inclined feature writer or public relations consultant. Herein are my directions to you for writing such an article. (The lessons to the rest of us are implicit in the piece.)

First, pick someone about whom something has already been written, if possible, since it's usually easier to write about a celebrity candidate who is already at the outer edges of the media hall of mirrors than to introduce someone absolutely new. (There is a danger, however, if your candidate is *too* well known for having accomplished something substantial; his or her image may not be sufficiently plastic.) Next, grossly exaggerate the importance of the person in question. Multiply, exponentiate, magnify. Claim that he or she is in the forefront of this movement, the epitome of that trend, the acknowledged leader of one significant development or other, or the possessor of some unique complex of physical and psychological traits. This isn't as difficult as it sounds since there are indefinitely many dimensions or combinations of dimensions along which we can measure people, and almost none of them—no matter how heinous, unethical, or ludicrous—is considered disqualifying anymore.

To establish lineage and achieve historic resonance (as much, at least, as our present-minded culture will allow), it is important to compare your celebrity-to-be with some luminary to whom he or she is vaguely and coincidentally similar, hence the reference to Woody Allen in the title of this segment. Again for mathematical reasons, this is surprisingly easy to do at all levels of celebrity, especially if you have access to clever advertising people or you're adept at searching Nexis or other large databases. Recall from section 2 the ease with which links may be established between almost any two people.

After you come up with some heroic parallels for your star-to-be, the formula requires an engaging title. Try something punning and marginally learned like DREISER'S ELECTRIC GUITAR MEETS LITTLE RESISTANCE, CARRIES BIG CHARGE or NO SKULDUGGERY FOR ANTHRO-POLOGIST JENKINS. Resume the hyperbole with more exclamations (remember that repetition is a virtue) about the new celebrity's previous accomplishments and renown among the cognoscenti. Be sure to quote the subject extensively on his or her own fame. The liar's paradox in logic derives from the fact that the statement "I am lying" is true if and only if it's false. The celebrity's paradox derives from the fact that the statement "I am famous" is true if and only if it's heard by a sufficient number of people.

To help spread this message of self-proclaimed fame, find a few back-scratching testimonials from people who know your subject and whose lives and careers are directly or indirectly bound to his or hers. (In mathematics this is known as log[arithm]rolling.) Again, this is no trick because of the unappreciated degree of connectedness among almost any pair of people. Within a given field, the degree of connectedness approaches the incestuous.

Finally, throw in some extraneous numbers and statistics and a poll result, sales figure, or stray journal article to lend the piece some gravitas. Sometimes the more irrelevant these numbers are, the better. Add a few cute turns of phrase, edit, polish, and, voilà: a "new celebrity" profile. And fame, whatever its provenance, naturally leads to more fame and, eventually, to riches in our increasingly self-referential, winner-takes-all economy. Q.E.D. Holy Euclid, forgive me.

Your assignment: Write such a piece about someone you know. Other mathematical approaches you may use can be found in the next segment.

Tsongkerclintkinbro Wins

Everybody's Got an Angle

With primaries, caucuses, state conventions, a glut of rules on eligibility, incessant polling, media watches, and reams of punditry, the Democratic candidates' spin doctors have ample material from which to fashion an argument that their man is the front runner.

Further evidence for this came over the weekend in the balloting that took place in Nebrarkamassacalowa.

The fifty-five voting members in this little-known state's caucus ranked the five candidates as follows:

Eighteen members preferred Tsongas to Kerrey to Harkin to Brown to Clinton. Twelve members preferred Clinton to Harkin to Kerrey to Brown to Tsongas. Ten members preferred Brown to Clinton to Harkin to Kerrey to Tsongas. Nine members preferred Kerrey to Brown to Harkin to Clinton to Tsongas. Four members preferred Harkin to Clinton to Kerrey to Brown to Tsongas. And two members preferred Harkin to Brown to Kerrey to Clinton to Tsongas (see diagram).

Tsongas supporters stolidly argued that the plurality method, whereby the candidate with the most first-place votes wins, should be used. With this method and eighteen first-place votes, Tsongas wins easily.

Ever alert for a comeback, Clinton supporters argued that there

This op-ed article appeared in the March 2, 1992, issue of the *New York Times*. At the time there were still five viable candidates in the race for the Democratic presidential nomination. The piece illustrates the utility of having a mathematician on one's campaign staff.

Nebrarkamassacalowans' Preferences

	Number of Delegates					
	18	12	10	9	4	2
First Choice	T	C	B	K	H	H
Second Choice	K	H	C	B	C	B
Third Choice	H	K	H	H	K	K
Fourth Choice	B	B	K	C	B	C
Fifth Choice	C	T	T	T	T	T

T = Tsongas C = Clinton B = Brown
K = Kerrey H = Harkin

should be a runoff between the two candidates receiving the most first-place votes. Clinton handily beats Tsongas in such a runoff (eighteen members preferring Tsongas to Clinton, but thirty-seven preferring Clinton to Tsongas).

Brown's people had to be a little more ethereal to come up with a method under which Brown would come out on top. Their suggestion: The candidate with the fewest first-place votes (Harkin in this case) should be eliminated first; then the first-place preferences for the others should be adjusted (still eighteen for Tsongas, now sixteen for Clinton, now twelve for Brown, still nine for Kerrey). Next, the candidate among the remaining four having the fewest first-place votes (Kerrey in this case) should be eliminated and the first-place preferences for the remaining candidates adjusted. (Brown now has twenty-one first-place votes.) This procedure of winnowing the candidates by removing at each stage the one with the fewest first-place votes is continued. Using this method, Brown is the winner.

Kerrey's campaign manager remonstrated that more attention should be paid to the overall rankings, not just to the top preferences. He argued that if first-place votes are each accorded 5 points, second place votes 4 points, third place 3 points, fourth place 2 points, and last place 1 point, then associated with each candidate will be a number that will accurately reflect that candidate's support. Since Kerrey's count of 191 is higher than anyone else's, using this method he wins.

Finally, Harkin, being a more macho sort, responded that only man-to-man contests should count and that, pit against any of the other four candidates in a two-person race, he comes out the winner. For example, he beats Kerrey twenty-eight votes to twenty-seven, and Clinton thirty-three votes to twenty-two. Harkin claims that he therefore deserves to be the overall winner.

All the numbers concocted here (and manipulated by drawing on the work of the American mathematicians William F. Lucas and Joseph Malkevitch and the eighteenth-century philosophers Jean-Charles de Borda and the Marquis de Condorcet) were obviously intended to show how the choice of a voting method can sometimes determine the winner. Although such anomalies don't always result, every method of voting is subject to them.

In listening to the candidates' arguments and weighing them, Nebrarkamassacalowans recalled the advice that the old lawyer gave his protégé: "When the law's on your side, pound the law. When the facts are on your side, pound the facts. And when neither is on your side, pound the table."

(What types of campaigns and candidacies are benefited or hurt by each of the above or other methods of tallying? Candidates from geographically large states might learn from Senator Harkin's supporters, who jokingly suggested during the primaries that their man was leading. In winning Minnesota, Iowa, and Montana, he had captured the largest land mass of any of the contenders.)

Florida Dentist Accused of
Intentionally Spreading AIDS

Rumors, Self-Fulfilling Prophecies, and
National Obsessions

Everyone's heard of the Florida dentist who infected, possibly intentionally, six of his patients with the AIDS virus. An investigation of the story by Stephen Barr in *Lear's* magazine and an excerpt in a *New York Times* op-ed in March 1994 revealed a number of lacunae in the case against the dentist. It appears that some, if not all, of the victims had other risk factors that might have exposed them. And the test used to determine whether two strains of the virus are the same is still controversial and not widely accepted. Furthermore, the rate of HIV infection among the dentist's many patients was only slightly higher than the overall rate in the two counties where the patients lived. The dentist himself may have been victim of the irrational fear induced by the AIDS epidemic.

The philosopher Daniel C. Dennett, in his book *Consciousness Explained,* describes a party game that I think provides a somewhat unusual slant on this and other cases in which there is an element of self-fulfilling prophecy. A familiar variant of the game requires that one try to determine by means of yes or no questions an arbitrary number between one and one million. Why, incidentally, do twenty questions always suffice for this?* In Dennett's more interesting game

---

*Each question can cut the number of possibilities in half. Thus the first question–Is it less than or equal to 500,000?–leaves $1,000,000(^1/_2)$ possibilities, two questions leave $1,000,000(^1/_2)^2$ possibilities, and twenty questions leave $1,000,000(^1/_2)^{20}$ possibilities. The final number of possibilities is less than one, so the number is determined. To locate a number between one and one billion requires only thirty questions, since $1,000,000,000(^1/_2)^{30}$ is less than one.

(which you might want to play if you have a friend you'd like to shed), one person is selected from the group. He will be asked to leave the room and told that, in his absence, one of the other party-goers will relate a recent dream. When the person returns to the room, he will attempt, through a sequence of yes or no questions directed to the group, to accomplish two things: reconstruct the dream and identify whose dream it was.

The punch line is that no one has related any dream. The revel-ers are instructed to respond either yes or no to the victim's ques-tions according to some arbitrary rule; Dennett suggests having the answers be determined by whether the last letter in the last word of the question is from the beginning or the end of the alphabet. Any rule will do, however, and may be supplemented by a noncontradic-tion clause stipulating that no answer directly contradict an earlier one.

The surprising result is that the victim, impelled by his own obsessions, often constructs an outlandish and obscene dream in response to the random answers he elicits. He may also think he knows whose dream it was, but then the ruse is revealed to him. Technically the dream has no author, but in a sense the victim him-self is. His preoccupations dictate his questions, which, even if answered negatively at first, will frequently receive a positive response in later reformulations. These positive responses are then pursued.

There is a body of experiment that seems to support the thesis that dreams and hallucinations can be explained in part by a variant of this party game. In both phenomena, a person's hypothesis-gener-ating ability is intact, but the ability to test or falsify these hypothe-ses is impaired by drugs, sensory deprivation, or unconsciousness. The result is a more or less random sequence of "answers" to the questions posed implicitly during the dream or hallucination. Unen-cumbered by any critical reality checks, the dreamer or hallucinator can fashion his or her own construction from this random set of answers.

A similar argument helps clarify why inane I Ching sayings or ambiguous horoscopes seem to many to be so apt. Their aptness is self-provided. In effect, their cryptic obscurity provides a random set of "answers" that the devotee fabricates into something seemingly appropriate and useful. With perhaps a bit more justification, psychologists count on the amorphousness of Rorschach blots to elicit evidence of a person's core concerns.

At the risk of stretching the point, I suppose that a somewhat similar phenomenon takes place for larger groups of people as well. Societies do not possess minds, of course, but in times of crisis—war, stock frenzy, pestilence, riot—they do develop a primitive sort of cohesiveness, a quasi consciousness approaching perhaps that of a very retarded person in a deep, drug-induced stupor. Because of the stress it endures, such a society will have indefinite fears, hopes, or anxieties, and its contact with reality will be tenuous. News reporting during a war or other crisis tends to be, for a variety of reasons, deplorably shoddy. (Recall the coverage of the Gulf War or during the early days of the AIDS crisis.) What news the society does get is vague and generic, allowing ample room for the societal analogue of dreams and hallucinations to develop. Societies without a free press and a literate population are especially vulnerable. A recent example is Guatemala, where peasants have seriously injured Western women they thought were kidnaping their babies. Another is Rwanda, where terrifying rumors and radio broadcasts inflamed and exacerbated an already nightmarish situation.

Ambiguity, randomness, and lack of information in response to obsessive questions and concerns can, on a group level, breed delusions and mirages in the same way that the party game makes an individual concoct his own chimerical fantasy. Informative, skeptical, fastidious reporting is most needed when it's least likely to be forthcoming.

Interlude: Selves, Heroes, and Dissociation

Many of the segments herein involve the notion of self. Perhaps our excessive concern with this problematic concept stems from the nagging suspicion that it is somewhat passé in this age of the planetary megalopolis. It's not the uncertainty that we will achieve personal fulfillment (whatever that might entail) or the belief that huge government and commercial databases imperil our privacy (as they do), but rather the frightening idea that we're easily replaceable and that the primary reality is the culture, the society, the weblike leviathan whose "mind" is reflected in the daily newspaper and in the electronic media. McLuhanesque insights about the disconnected collage that the media present and with which we resonate remain accurate. Reading newspapers and magazines, switching channels on cable TV, and floating in the all-enveloping cyberspace certainly foster a modernist fragmentation.

A little of this and a little of that. Meaningless juxtapositions and coincidences replace conventional narratives and contribute to our dissociation. The comedian George Carlin once listed six reasons for doing something or other: 1, b, III, four, E, and vi; this is the notational system of the media. "World trouble spots" change in quick succession, and a sequence of societal problems and dysfunctions is always before us. We're constantly exposed to individuals who are curiously unindividual, icons for some aggrieved group or other. The magnification of public personalities also makes it easy to think of the self as merely a particular embodiment of social reality. Personality or self seems to be a nominal entity, like that possessed by, say, the Philadelphia Phillies; it consists of many interacting and sometimes contrary subparts that don't really cohere, that have no center.

Increasingly, the basic units of society are ethnic groups, corporations, organizations of various kinds. Corresponding to this growth in the importance of the organization is a diminution in that of the individual. It's become more and more apparent that along most dimensions that we care to examine, people are quite ordinary, average, normal. As biographies, obituaries, exposés, and personal experience all attest, this is true of even the most remarkable personages outside their areas of expertise or accomplishment. I used to be surprised and even depressed when I met someone I had admired and discovered him or her to be a jerk or, perhaps worse, rather common. Now I half expect this reaction and even find it a little reassuring, maybe even uplifting. Isn't it amazing that somebody like that could produce such and such?

What holds true of the notion of self is doubly so for that of hero. The skeptical intrusiveness of the media (which I ambivalently applaud) makes it very difficult to believe in great heroes or great villains. The flawlessly ideal or demonically hateful image that is required is harder to maintain (except in times of war or other crisis). The situation and ambient society play a larger role than in the past. People may simply all be more or less people size–a discouraging realization for aspiring heroes and heroines as, I suspect, many of us still are.

The question looms: What is a self in this strange age of dissociation and connection?* Working as a member of an interdependent organization, relying on technology and organized industry, being exposed to the disjointed cacophony of the mass media–all these things tend to overwhelm us. Our increasingly integrated and regimented society undermines our sense of self, sometimes with

---

*The computer scientist Marvin Minsky's answer in *The Society of Mind* is that the self is a parliament of little, semiautonomous processes whose interactions add up in some strange, poorly understood way to an idiosyncratic whole, a whole that may, under the appropriate circumstances, answer a rhetorical question about its own definition in a footnote. Whatever the details of this definition, evolution has ensured that we take the self seriously. Doing so helps keep us alive and preserves our genes from extinction; thoroughgoing nihilists aren't very prolific.

totalitarian earnestness, more often with relentless chirpiness. Out of what, then, is one to fashion and maintain a self? Out of the trivial quirks that characterize each of us, out of fleeting relationships that differ hardly at all from millions of others, out of making a minuscule contribution to the technology of the mass society? Probably not. Out of art, science, family, friends, and love? More likely. More than uniqueness is certainly required; we're all unique in the picayune sense in which 2,452,983,448 and 3,887,119,932 are unique. We also need coherence, a point of view, complexity.

Achieving personal integration and a sense of self is for the benefit of ourselves and those we're close to. Doing so will not make us obviously distinguishable from others nor win us any prizes. Related to this point on distinguishability is an idea I've had concerning vanity license plates. While I've never been tempted to dream up some bit of capsule wisdom to adorn my bumper, I do like the idea of generating some random sequence of letters and numbers and requesting that this be my vanity plate. I'd have the absurd satisfaction of knowing that my ordinary-looking plate was really a vanity plate. Maybe this is the closest we can come to heroism.

Candidates Contradict Each Other's Denials

Self-Reference, Intentions, and the News

News reports in general, and celebrity coverage in particular, are becoming ever more self-referential. What's reported is frequently not some state of affairs out in the world but rather some other report that is itself an account of yet another piece, each article in the sequence sprinkled liberally with the cascading reactions of players and observers alike. These reports can reinforce one another and result in a sudden surge of concern over some issue that, like a house of cards, just as rapidly fades from view. A Yankelovich study of what Americans over the last fifteen years have said is "the greatest problem facing the country today" shows spiky peaks of anxiety about foreign policy, the economy, drugs, and crime alternating with wide troughs of indifference toward these same matters.

Increasingly common ingredients in news stories are statistics gathered from Nexis, Lexis, and other databases reporting on the number of articles dealing with this or that matter of concern. These numbers are similarly volatile. Whether in daily life or in the newspaper, however, obsession with who said what, with nth hand reactions to what has been said, with how many times a topic is mentioned, and with continual self-reference is generally not healthy. Contrariwise, neither is suppression of self and its role in a story's origin.

Before turning to these topical issues, let me digress a bit to discuss the different associations that come to mind when a mathematician thinks of self-reference.* I'm reminded, for example, of a puzzle called the paradox of the preface. It is an author's innocuous remark in the foreword to an article or book modestly, yet inconsistently, admitting that although he or she stands behind all the claims made in the piece, it undoubtedly contains some false ones. Another example occurred in my earlier discussion of simplistic brevity, wherein I mentioned the self-referentially paradoxical slogan: No More Slogans.

Related logical difficulties arise in many different guises. The classical liar paradox referred to earlier in this section can result, for example, if Senator S simply announces that everything he says is false. If his statement is true, then it's false; and if it's false, then it's true. More complicated occurrences involving two or more parties can also easily arise. If candidate X says that candidate Y's comments about the crime bill are false, there is nothing paradoxical about her statement. If candidate Y says that candidate X's remarks about the bill are true, there is nothing paradoxical about this statement either. But if we combine these two statements, we have a paradox. It's not too hard to imagine a collection of such comments from a variety of people, each individually plausible, yet leading to an equally potent paradox. Such is likely to be the case with incestuous media reporting and political spin-doctoring.

A diverting puzzle concerns the reporter who knows that his source either always tells the truth or always lies, but forgets which. The reporter wants to know whether Senator S is involved in a certain scandal and can ask his source, who knows the answer, only one yes or no question. What should it be?† The reporter faces a more

---

*Philosophy professors tell jokes about how one can always identify a philosopher: He is the one at the conference on the criminal justice system, for example, who delivers a paper on the prisoner's dilemma (an idealized conundrum about the trade-off between cooperation and self-interest). The point of the joke, which generally amuses few people except philosophers, is that the theoretical puzzle is, despite its name, far removed from the practical concerns of the conference. Some of the ideas in this book, and especially in the next two segments, may strike the reader as similarly remote from traditional journalistic criticism. Suffering perhaps from the same professional myopia as philosophers, I nevertheless have confidence in the power and relevance of abstraction.

†Is it the case that the two statements—*You are a truth-teller* and *The senator is involved in this scandal*—are either both true or both false? The remarkable thing about this question is that

difficult problem in a puzzle adapted from one by the logician Ray Smullyan. He again wants to know whether Senator S is implicated in the scandal, but this time he has three knowledgeable informants, A, B, and C. One is a truth-teller, one a liar, and one a normal person who sometimes lies and sometimes tells the truth. The reporter doesn't know which is which, but he can ask *two* yes or no questions, each directed to a single informant, to determine Senator S's culpability. What questions should he ask and of whom should he ask them?* The moral of the story is that complete liars can be as informative as truth-tellers. The problem is with those pesky critters who sometimes lie and sometimes tell the truth.

Interestingly, self-reference makes an assumption of complete knowledge untenable. Imagine a supercomputer, the Delphic-Cray-1A, into which has been programmed the most complete and up-to-date scientific knowledge, the initial conditions of all particles, and sophisticated mathematical techniques and formulas. Assume further that 1A answers only yes or no questions, and that its output device is constructed in such a way that a yes answer turns off an attached lightbulb if it's on, and a no answer turns on said lightbulb if it's off. If we ask this impressive machine something about the external world, let us assume for argument's sake that it responds impeccably. If we ask it whether the lightbulb will be on in one hour, however, 1A is

---

both truth-teller and liar will answer yes if the senator is involved and no if he is not. Think about it. This is a general principle. If you want to know whether any statement S is true and your interlocutor is a liar or a truth-teller, ask him if the two statements–You are a truth-teller and statement S–are either both true or both false. You can trust the answer.

*Since truth-tellers and liars are preferable to normal people, the goal here is to use one of our questions to find someone who isn't normal. Once we've located him, we've reduced the problem to the one in the previous note. Thus the first question should be directed toward A and it should be: Is it the case that the following two statements–*You are a truth-teller* and B *is normal*–are either both true or both false? Assume A answers yes. If A is a truth-teller or a liar, then we know we can trust the answer, and B must be normal and hence C not normal. If A is not a truth-teller or a liar, then he must be normal and again we conclude that C is not normal. Either way, a yes answer means that C is not normal. On the other hand, if A answers no and is a truth-teller or a liar, then we can trust his answer and conclude that B is not normal. If A is not a truth-teller or a liar, then again we know that B is not normal since A is. Either way, a no means that B is not normal. If we get a yes, we ask C the next question; if we get a no, we ask B. The second question is the one posed in the previous note.

stumped. Quizzing 1A about itself immediately induces a giddy oscillation: if yes, then no; if no, then yes. The computer cannot isolate and analyze such questions since they engage the whole machine.

There are various approaches (which you may want to skip) that are useful in clarifying such logical snarls. The oldest, developed by Bertrand Russell and refined by the logician Alfred Tarski, is the notion of a statement's logical level or order. Statements about the world ("The sod is a toupee for my yard") are termed first-order statements, while statements about first-order statements ("His remark on sod is typical of his obsession with hair") are termed second-order statements. Third-order statements are about second-order statements, and so on for fourth-order, fifth-order, and other meta-level statements. In this way, for example, Senator S's announcement that everything he says is false would be taken to be a second-order statement describing his first-order statements only, and the paradox would be averted.

Another method of avoiding self-referential paradoxes is attributable to the logician Saul Kripke. It does not assign a fixed order to statements, but allows them to attain their order naturally depending on what other statements have been made and on the facts of the situation. The truth or falsity of statements is decided in a gradual, step-by-step manner with some self-referential statements remaining indeterminate.

More broadly construed, self-reference underlies our fundamental understanding of all social processes. A long tradition in sociology dating back to Max Weber maintains that identification with others is essential to understanding social regularities since they depend as much on human rules (red lights meaning stop, for example) as on scientific principles. The problem is that these rules of human behavior are notoriously vulnerable to self-fulfillment, and the distinction between empirical laws and tautological conventions is much fuzzier than it is in the physical sciences. Nevertheless, social discourse *requires* that one refer to oneself, identify with others, and internalize social practices. No matter how independent and sovereign we think we are, we must take the actions and statements of others into account before we do or say something. And in reading the news, we should be cognizant that the meta-level aspects of a story can have a significant effect on our interpretation of it.

Let me illustrate how we can be bound up in self-referential tangles willy-nilly. In a perverse mood, I informed one of my classes early in the semester of a new rule: Anyone who checked a box on the exam sheet would have an extra ten points added to his or her exam score unless the number of students checking the box constituted more than half the class. If more than half the class checked the box, those checking it would have ten points deducted from their exam scores. Since the students were concerned about their rank in the class, even those who didn't wish to gamble had to take the actions of their fellow students into account in deciding whether to check the box. What happened is that progressively more students checked the box on the exam sheets as the semester wore on, until on one exam more than half the class checked the box and these students were penalized ten points. Very few students checked the box on exams thereafter.

An anecdote with a similar lesson comes from Professor Martin Shubik, who auctioned off one dollar to students in his class at Yale. Bidding took place at 5-cent intervals. The highest bidder got the dollar, of course, but the second highest bidder was required to pay his bid as well. For example, if the highest bid was 50 cents and you were second highest at 45 cents, the leader stood to make 50 cents on the deal and you would lose 45 cents were the bidding to stop there. You'd have an incentive to up your bid to at least 55 cents, but then the other bidder would have an even bigger incentive to raise her bid as well. In this way a one-dollar bill was successfully auctioned off for two, three, four, or more dollars.

Such interactions are, of course, much more complicated in the hurly-burly world of politics and business, where the rules of interaction are not known, much less explicitly stated. And there is yet another issue of self-reference in these latter domains. Newspaper conventions sometimes make it difficult for readers to ascertain some of the crucial meta-level aspects of a story because reporters aren't generally allowed to refer to a newsmaker's intentions or to their own involvement in the story.

For instance, I've been involved in demonstrations that were pretty languid affairs until the cameras and reporters showed up, at which time people abruptly began gesticulating wildly and spouting angry rhetoric. This significant and somewhat humorous fact was never reported in the news stories about the demonstrations. Likewise, candidates' proposals or corporations' press releases are usually reported in the terms of the candidates or corporations themselves—even when it's clear to the reporters present that these announcements are carefully staged and timed performances intended to exploit the journalistic convention of reportorial self-effacement.

A related example in which I had a peripheral involvement concerns a politically liberal professor who announced at a well-attended faculty meeting that he was a racist. He intended by this remark to show that almost everyone, even he, harbored some racist attitudes, and that a proposed course to be taught on racial awareness should therefore be required of all students. Realizing that their colleague was engaging in some holier-than-thou posturing, other faculty members rejoined that he was only boasting about his supposed racism in order to highlight his keen sensitivity to racial issues. Later some black students, most of whom were in favor of the proposed course, heard literal media accounts of the meeting—PROFESSOR BOASTS OF RACISM—and concluded, reasonably enough, that some professor had publicly boasted about his racism. In response, they organized a protest demonstration.

More blatant is the headline POPE DISRUPTS BAR MITZVAH, which might be accurate but is similarly misleading; perhaps the pope's motorcade through the city simply upset the normal traffic patterns and thereby prevented guests from reaching the synagogue.

It is perhaps needless to repeat that such literal reports can be deceptive. Wittgenstein asked: What remains if I subtract my arm's going up from my raising it? The answer, of course, is the agent's intention. The newsmaker's intentions frequently are a part of the story, and so, occasionally, is the reporter's role in it.

Journalistic neutrality should not preclude, for example, a reporter's writing that a given story on some government official was strongly urged upon him or her during an airport meeting by someone sympathetic (or antagonistic) to that official. We're clueless enough about how and by whom issues are decided without critical information about the reporter's own role in the story being omitted as a matter of course. The real story is often not some statement X put out by Y, but that Y said X with a specific intention, or that Z wanted Y to be linked with X, or that W used Y's desire to link Y with X to advance his own goals, and that reporter R is associated with Y, Z, or W, or knows of these or other associations, or even is Y, Z, or W. This brings me to the topic of complexity and the news.

Special Investigator Says Full Story Not Told

Compressibility and the Complexity Horizon

The convolution of people's interacting expectations (I thought that she had told you that we were considering his invitation to . . .) and the delicate web of human organization that they constitute bring me to the "complexity horizon"—that limit or edge beyond which social laws, events, and regularities are so complex as to be unfathomable, seemingly random. Applied loosely and casually, the term is useful in referring to discriminations that are impossibly subtle for a given group of people at a given point in time.

Thus, if a set of dense tomes recounting the full history of, say, the cold war were presented to people who were capable of grasping only the notion of a fistfight, the conflict's full history could be said to be beyond their complexity horizon. An in-depth account of the savings-and-loan scandal would be beyond the complexity horizon of those whose economic experience consisted entirely of making change. A few angstroms' difference in length to tribesmen who measured objects with tree branches would likewise be beyond theirs.*

Happily for us, most laws and regularities essential for life are not beyond our complexity horizon and are compressible enough to be grasped. Compression and simplification certainly seem essential

---

*Switching to a different tribe, I note that no mathematician can read with understanding more than a handful of the hundreds of mathematics journals currently being published. In this sense, most research mathematics is beyond every mathematician's complexity horizon.

to short-term political success. James Fallows in a *Los Angeles Times* story has even suggested that presidents who overanalyze fail, while those who oversimplify succeed.

These informal notions of complexity and compressibility are of some relevance to news stories. The primary purpose of the latter, after all, is to compress the complex set of facts surrounding an event into something comprehensible. There are unresolved legal issues about how this can be accomplished. *New Yorker* writer Janet Malcolm's practice of compression, which she utilized in her profile of the psychoanalyst Jeffrey Masson, is a case in point. Apparently, Ms. Malcolm included quotations from a series of different encounters with Dr. Masson in a single sentence, thereby heightening their dramatic impact. Although the following brief excursion into a formalization of these notions doesn't settle this legal matter, it does allow for a revealing look at some other limitations of compression.

Assuming you had some interest in doing so, how would you describe these sequences to an acquaintance who couldn't see them?

1.   0 0 1 0 0 1 0 0 1 0 0 1 0 0 1 0 0 1 0 0 1 0 0 1 0 . . .
2.   0 1 0 1 1 0 1 0 1 0 1 1 0 1 0 1 1 0 1 0 1 0 1 0 1 . . .
3.   1 0 0 0 1 0 1 1 0 1 1 0 1 1 0 0 0 1 0 1 0 1 1 0 0 . . .

Clearly, sequence number one is the simplest, being merely a repetition of two 0's and a 1. Sequence number two has some regularity to it–a single 0 alternating sometimes with a 1, sometimes with two 1's–while sequence number three is the most difficult to describe, since it doesn't seem to evince any pattern at all. Observe that the precise meaning of the final ellipses in the first sequence is clear; it is less clear in the second sequence, and not at all clear in the third. Despite this ambiguity, let's assume that these sequences are each a trillion bits long (a *bit* is a 0 or a 1) and continue on in the same way.

We can follow the example of the computer scientist Gregory Chaitin and the Russian mathematician A. N. Kolmogorov and define the complexity of a sequence of 0's and 1's to be the length of the shortest computer program that will generate (or print out) the sequence in question.

Note that a program that prints out the first sequence can consist of the following simple prescription: Print two 0's, then a 1, and repeat. The information in this trillion-bit sequence can be compressed into a very short program that generates the sequence. Thus, the complexity of this first sequence may be only about a thousand bits, or however long the shortest program generating it turns out to be. (This depends to some extent on the computer language used to write the program.)

A program that generates the second sequence would be a translation of the following: Print a 0 followed either by a single 1 or two 1's, the pattern of the intervening 1's being one, two, one, one, two, one, two, one, and so on. If this pattern persists, any program that prints out the sequence would have to be quite long so as to fully specify the "and so on" pattern of the intervening 1's. Nevertheless, because of the regular alternation of 0's and 1's, the information in the trillion-bit sequence can be compressed into a program that is somewhat shorter than the trillion-bit sequence that it generates. Thus the complexity of this second sequence might be only half a trillion bits, or whatever the length of the shortest program generating it is.

With the third sequence, the most common sort, the situation is different. The sequence, let us assume, remains so disorderly throughout its trillion-bit length that no program that generated it would be any shorter than the sequence itself: The information in the sequence is incompressible. All any program can do in this case is to list dumbly the bits in the sequence: Print 1, then 0, then 0, then 0, then 1, then 0, then 1, then. . . . Such a program will be at least as long as the sequence it generates, and thus the third sequence has a complexity of approximately a trillion. How is it that mnemonic devices that lengthen what must be remembered are sometimes useful?*

---

*If the mnemonic device merely attaches an extraneous story to every digit to be memorized, it does no good. For example, if a woman were to memorize a telephone number, say 253-3784, by recalling that her best friend has 2 children, her doctor has 5, her college roommate 3, her neighbor on the left has 3 dogs, the one on the right 7 cats, her older brother has 8 children if you counted those of his ex-wives, and she herself is one of 4 children, she would gain nothing. This "program" is longer and more convoluted than the one it was designed to help remember. Sometimes, however, when a program is intimately linked with a list or a story that one already knows well, its length is only apparent, and recalling one or two facts might be sufficient to recall the number.

A sequence like the third one, which requires a program as long as itself to be generated, is said to be *random*. Random sequences evince no pattern, regularity, or order, and the programs that generate them can do nothing more than direct that they be copied: print 1 0 0 0 1 0 1 1 0 1 1.... These programs cannot be condensed or abbreviated.

In some ways, sequences like the second are the most interesting since, like living things, they display elements of both order and randomness. Their complexity is less than their length, but not so small as to be completely ordered or so large as to be random. The first sequence may perhaps be compared to a precious stone or a salt crystal in its regularity, while the third is comparable to a haze of gas molecules or a succession of dice rolls in its randomness. The analogue of the second might be something like a rose or, less poetically, a newspaper, which manifests both order and randomness among its parts. (A quote from the French poet Paul Valéry is apt here: "Two dangers threaten the world—order and disorder.")

These comparisons are more than mere metaphors. The reason is that most phenomena can be described via some code, and any such code—whether it be the molecular alphabet of amino acids assembled into the stories told by DNA molecules, the strings of symbols that represent the operations of an idealized computer, or the English alphabet assembled into newspaper articles—can be digitalized and reduced to sequences of 0's and 1's. Expressed in their respective codes, both DNA and news stories are sequences like the second one, evincing order and compressibility as well as complexity and randomness. Similarly, most melodies lie between repetitive beats and formless static (analogous, respectively, to sequences like the first and the third).

This notion of complexity is called *algorithmic* complexity because it gives one the length of the shortest program (algorithm, recipe) needed to generate a given sequence. Using it, Chaitin proved that the output of any computer (or, in fact, any human artifact) is limited in the complexity of the sequences it can generate. Specifically, no computer can produce sequences that are more complex than it itself is. Imagine if a computer (whose operation can be coded up into a sequence C of 0's and 1's) were able to generate a random sequence S more complex than it itself. Then to generate S we could merely write a relatively short program to generate C, the sequence of lower complexity that represents the computer, which could in turn generate S. This would mean that S, contrary to assumption, is not more complex than C; neither would it be random, since it would have been generated by a program shorter than itself.

Germane to these matters is the so-called Berry paradox: "Find the smallest whole number that requires, in order to be specified, more words than there are in this sentence." The number of hairs on my head, the number of different states of the Rubik's Cube, and the speed of light in millimeters per century each specify, using fewer than the number of words in the given sentence (twenty), some particular whole number. No problem so far. The paradoxical nature of the Berry task becomes apparent, however, when we note that the sentence specifies a particular whole number that, by definition, it contains too few words to specify.

Chaitin's theorem, which is a generalization of Godel's incompleteness theorem,* is not a paradox, although it is a strange and deep mathematical result. Very loosely paraphrased, it states that every computer, every formalizable system, and every human production is limited; there are always sequences that are too complex to be generated, outcomes too complex to be predicted, and events too dense to be compressed. Sequences, outcomes, and events that are this complex can be said to be beyond the complexity horizon of the information-processing system in question.

This would be an academic exercise in logic were it not for the fact (at least I believe it's a fact) that increasingly many aspects of modern life have become so intricately tangled and convolutedly complicated as to be incompressible and beyond our complexity horizon. We can't grasp them. The most we can do is what one would do with sequence number three: Just wait and see and perhaps speculate

---

*The (first) incompleteness theorem states that any formal system of mathematics that includes a modicum of arithmetic is incomplete; there will always be true statements that will be neither provable nor disprovable within the system, no matter how elaborate it is.

about an underlying order. Once at or beyond our complexity horizon, we're relatively unconstrained by facts and may read into the amorphous swirl of data at hand largely what we wish (a bit like the man in the party game). There are seldom nice little recipes that predict more than a short distance into the future. In the public realm, often the best we can do is to stand by and see how events unfold.

Newspaper Circulation Down
Factoids on Tabloids

What is more self-referential than a newspaper story about news-papers? Variants of the above headline are common, but my bet is on the continuing vitality of the news business. Worldwide there are an estimated 60,000 newspapers, with a combined circulation of 500 million and a readership three times as large. In the United States there are approximately 2,600 newspapers, 1,800 of which are dailies, and the combined circulation is 75 million.

These numbers will be more vivid if we attach a few names to them. (I don't want to miss a chance to mention some newspaper names, marginally relevant though they may be. A residue of my newspaper fetish, no doubt.) Among the leading papers in the world are the *Toronto Globe and Mail* in Canada, *The Times* and *The Guardian* in England, the *Frankfurter Allgemeine* and the *Suesseutsche Zeitung* in Germany, *Le Monde* and *Le Figaro* in France, *Pravda* and *Izvestia* in Russia, *El Pais* in Spain, the *People's Daily* (*Renmin Ribao*) in China, the *Asahi* and *Mainichi* in Japan, *The Indian Express* and *The Times* in India, *Excelsior* in Mexico, *El Mercurio* in Chile, and *La Nación* in Argentina.

The *New York Times*, *Washington Post*, and *Los Angeles Times* are the Big Three of serious newspapers in the United States, followed in no particular order by the *Wall Street Journal*, *Philadelphia Inquirer*, *Boston Globe*, *Chicago Tribune*, *Atlanta Constitution*, *Miami Herald*, *Louisville Courier-Journal*, *USA Today*, and a dozen or two other solid papers.

I have seen purported circulation figures, but I have my doubts about them. The number of newspapers, for example, is clearly dependent on the definition used. Do we count community newsletters, various large organizations' publications, special interest publications? And how do we count the papers in a chain? In 1900 there were 8 chains in the United States, which together controlled twenty-seven papers. Now more than 150 chains control about 70 percent of all daily papers. (In the spirit of self-reference, I note that one of the biggest chains belongs to Rupert Murdoch, whose vast holdings include Basic Books, publisher of this book.) In any case, it's sometimes forgotten that what's important is keeping the number of independent news *sources* large, not necessarily the number of newspapers.

The figures on readership and circulation are also ambiguous. Although readership is often greater than circulation (because families, institutions, commuters, and others often share the same newspaper), the total number of people who read is smaller than the total number of readers reported. As is the case with airlines who claim X million passengers annually, newspaper readership (and circulation) claims are not for distinct human beings. If George reads three papers a day and takes four flights a year and Martha reads two papers and takes seven flights a year, together they constitute five readers and eleven passengers. (McDonald's doesn't make the same mistake, never claiming that 100 billion people have eaten their hamburgers.)

Whatever the numbers, the newspaper business is an indispensable one, and despite some declines in circulation (counterbalanced by significant increases during times of crisis), it is likely to remain a lucrative one as well. Although necessarily ceding coverage of the immediate breaking story to television, newspapers continue to cover events in greater depth than TV news, with background stories, engaging features, and knowledgeable analysis. But there's ample room for many different media. Seeing a news or sporting event on television, listening to a radio commentary about it, or eavesdropping on an on-line discussion makes me more rather than less eager to read about it in the paper the next day. More tie-ins with other media may, in fact, be the way to lure greater numbers of Generation X readers.

Computers, Faxes, Copiers Still Rare in Russia

Information and the Commissars

Without computers, faxes, modems, and copiers, it is much easier for comprehension of events and situations to be beyond our complexity horizons. The suggestion that the failure of command economies in the former Soviet Union and Eastern Europe might have been due as much to information-theoretic limitations as to political ones is not preposterous. Undoubtedly, the commissars found it increasingly burdensome to centrally coordinate exponentially proliferating data on things like supplies, foodstuffs, and parts. The predicament is hard to quantify, yet universal. Now that a laser printer can transform a personal computer into a publishing house or a type foundry, our ability to sort and retrieve information is lagging ever farther behind our ability to produce it.

As business reports and scientific research papers, newspapers and other periodicals, databases and electronic mail, textbooks and other books all increase rapidly, the number of interdependencies among them rises exponentially. New ways to link, classify, and order the traffic on the information highways of the future are necessary if we're to thread our way through the mounds of raw data strewn along them. Some understanding of basic mathematical and statistical ideas is necessary if we're to avoid the condition described by the computer scientist Jesse Shera's paraphrase of Coleridge: "Data, data everywhere, but not a thought to think."

If I may combine terms from disparate realms, the Jeffersonian model of many parallel processors is superior to the Stalinist model of one central processor.* We don't need a controlled media and party apparatchiks who will mechanically respond to government dicta; we need an independent press and free men and women who will have to make sense of the unforeseen complexities of the twenty-first century.

---

*Recall Jefferson's remark: "Were it left to me to decide whether we should have a government without newspapers, or newspapers without a government, I should not hesitate a moment to choose the latter."

Section 4

SCIENCE, MEDICINE, AND THE ENVIRONMENT

It ain't what you don't know that counts. It's what you know that ain't so.

–WILL ROGERS

In this section, I'll examine some of the mathematical aspects of science stories in the daily paper. Many such stories announce new studies or developments but fail to put the announcement into any sort of perspective. This is likely to be misleading, even when the story is technically accurate. As Bertrand Russell once wrote, it is sometimes necessary to choose between clarity and precision, and an enlightening clarity (without serious distortion) is to be preferred to an obfuscating precision, especially in the newspaper.

Some scientists don't see it this way, though; they act as if the newspaper were simply a professional journal with a very large circulation or some sort of public relations office for their laboratory or university. At the other extreme are those reporters who latch on to the most alarming scenarios consistent with the new results. The heroes are those whose knowledge of science is broad and detailed enough, and whose writing is sufficiently clear and engaging, to communicate the science to the general public effectively. Sometimes these people are first-class scientists, but more often, it seems, they're science journalists.

What follows is a discussion of the statistical pitfalls to avoid if you want to interpret news stories on health risks properly. Then there are segments on the near inevitability of pollution, the unpredictability of ecological trends, the appeal of pseudoscience, the dismal math scores of American students, and related matters.

Ranking Health Risks: Experts and Laymen Differ

The Dyscalculia Syndrome

Health statistics may be bad for our mental health. Inundated by too many of them, we tend to ignore them completely, to react to them emotionally, to accept them blithely, to disbelieve them closed-mindedly, or simply to misinterpret their significance. (The National Institute of Unchallengeable Statistics reports that 88.479 percent of us have one of these five reactions 5.613 times per day, leading to the 8,373,429 cases of dyscalculia recorded annually.) Since the injunction to show and not tell extends beyond classes on creative writing, I'll focus here on specific examples that illustrate successively the psychological, mathematical, and factual lapses that underlie many of our inappropriate reactions to statistics and on the ways in which newspapers foster these responses.

First, the psychological component of dyscalculia. In previous segments on cellular phones, military death tolls, and other topics, I have demonstrated the impact of the "availability error" on mathematical common sense. Anxiety, fear, and the force of a vivid anecdote can, for example, mask the mathematical distinction between the rate of incidence of some condition and the absolute number of its instances. In a country the size of the United States, an extremely rare condition that distresses, say, one in a million people will still

This segment appeared in the March 1994 issue of *Discover* magazine.

afflict 260 people. The bias termed "anchoring effects," also discussed in a previous segment, plays a role in the perception of health risks as well: People generally remain anchored to the first number they hear, whether accurate or not.

Even round numbers have a psychological appeal for us—specifically, multiples of ten. It's been maintained for years, for example, that 10 percent of Americans are homosexual, that we each use only about 10 percent of our brain capacity, and that the condom failure rate is 10 percent. Such statistics are artifacts, I presume, of our decimal system; in a base 12 system, we'd no doubt have many $8^1/3$ percent statistics. These numbers, although few of us understand precisely what they mean, once accepted, become resistant to significant revision.

When health risks involve large numbers and fall outside our daily frame of reference or beyond our personal control, we're more likely to be swayed by psychological factors and misjudge the real hazard. For example, drugs are an undeniable scourge, but the biggest killers among them are tobacco (400,000 annually) and alcohol (90,000 annually), not cocaine (8,000) or heroin (6,000), which nevertheless have more emotional impact and induce more alarm. As the physicist H. W. Lewis has written, nuclear power plants are feared by most Americans, yet the prosaic problem of lead in old paint and old pipes has caused far more harm. Likewise, the famed California biochemist Bruce Ames has estimated that we ingest 10,000 times as much natural as man-made pesticides: the estragol in basil, the hydrazines in mushrooms, the aflatoxins in peanuts, and so on. But no one rides around with a bumper sticker reading *No Peanuts*. People worry that electromagnetic fields, which have increased by a factor of ten in the last fifty years, have caused an increase in leukemia rates, which in fact have risen slightly, if at all, in that time.

Such unfounded, or at least excessive, private concerns are not without public consequences. They have led, for example, to passage of such benighted legislation as the Delaney Clause of the FDA Act of 1958, which requires that "no food additive shall be deemed safe if it is found . . . to induce cancer in man or animal." Since its adoption, the law's scope has been widened to include pesticides and other pollutants–but no minimum allowed level has ever been specified. And in the subsequent expensive, time-consuming, and impossible effort to reduce minuscule risks to zero, little money or energy is left for substantial, albeit more banal, hazards.

Next, there are the statistical niceties, more to the mathematical side of the psychology-mathematics continuum. Consider Simpson's Paradox, which involves not the media's coverage of the O. J. spectacle but a particular, easily made arithmetical error: concluding that if you average several sets of numbers (or percentages) and then average these averages, the resulting number will be the average of all the numbers. Thus, if a study indicates that 36 percent of ethnic group A and 45 percent of ethnic group B improve from some treatment, and a second study indicates that 60 percent of group A and 65 percent of group B improve, it is tempting but incorrect to conclude that a higher percentage of group B improves. The first study might, for example, have included 100 members of group A and 1,000 members of group B, while in the second study these numbers might have been reversed. If so, how many of the 1,100 members of each ethnic group improved?*

Now consider the following different but related scenario, which suggests how a headline such as HALF OF SUFFERERS ARE LONG TERM might come about. Mr. X has suffered from a certain disease for years. In January, X and A are both treated for this disease. In February, A finds himself cured, but B contracts the disease. In March, B is better, but C reports for treatment, and so on. Poor Mr. X continues to suffer throughout the year. If one examines the record for any given month, however, one finds that 50 percent of the ailing people (always X and one other person) are chronic sufferers. Yet only $1/13$th of the people who suffer from this disease in any given year suffer from it for long.

An understanding of the mathematical notion of "conditional probability" is crucial to the proper interpreting of statistics. As men-

---

*Thirty-six percent of the 100 members of ethnic group A in the first study and 60 percent of the 1,000 members of A in the second study improved, for a total of 636 A members. Forty-five percent of the 1,000 members of ethnic group B in the first study and 65 percent of the 100 members of B in the second study improved, for a total of 515 B members.

tioned in an earlier segment, the probability of someone speaking English, given that (or on the condition that) he or she is an American citizen, is, let's assume, 95 percent. The conditional probability that someone is an American, given that he or she speaks English, is much less—say, 20 percent.

For a compelling application of conditional probability, consider this instance of Bayes's theorem You've taken a test for dread disease D (perhaps even dyscalculia), and your doctor has solemnly advised you that you've tested positive. How despondent should you be?

To see that cautious optimism may be appropriate, suppose there is a test for disease D that is 99 percent accurate in the following sense. If you have D, the test will be positive 99 percent of the time, and if you don't have it, the test will be negative 99 percent of the time. (For simplicity, I'm using the same percentage for both positive and negative tests.) Suppose further that .1 percent—one out of every thousand people—actually has this rare disease.

Let's now assume that 100,000 tests for D are administered (see diagram). Of these, how many will be positive? On the average, 100 of these 100,000 people (.1 percent of 100,000) will have D, and so, since 99 percent of these 100 will test positive, we will have, on average, 99 positive tests. Of the 99,900 healthy people, 1 percent will test positive, resulting in a total of approximately 999 positive

False Positives

	Sick	Healthy	
Test Positive	99	999	1,098
Test Negative	1	98,901	98,902
	100	99,900	100,000

Conditional probability that one has D given one's
tested positive is $99/1,098$, or a bit over 9%.

tests (1 percent of 99,900 is 999). Thus, of the total of 1,098 positive tests (999 + 99 = 1,098), most (999) are false positives, and so the conditional probability that you have D given that you tested positive is $^{99}/_{1,098}$, or a bit over 9 percent–and this for a test that was assumed to be 99 percent accurate!

To reiterate, the conditional probability that you test positive given that you have D is 99 percent, yet only 9 percent of those with positive tests will have D.

The whole panoply of statistical tests, estimates, and procedures gives rise to many mathematical nuances that may have practical consequences. Determining, for example, when clusters of a particular disease constitute evidence of something seriously awry or merely a coincidental clumping is not easy (especially for people who insist on reading significance into everything). And most people still don't realize that what's critical about a random sample is its absolute size, not its percentage of the population. Although it may seem counterintuitive, a random sample of 500 people taken from the entire U.S. population of 260 million is generally far more predictive of its population (has a smaller margin of error) than a random sample of 50 taken from a population of 2,600.

A more elementary widespread confusion is that between correlation and causation. Studies have shown repeatedly, for example, that children with longer arms reason better than those with shorter arms, but there is no causal connection here. Children with longer arms reason better because they're older! Consider a headline that invites us to infer a causal connection: BOTTLED WATER LINKED TO HEALTHIER BABIES. Without further evidence, this invitation should be refused, since affluent parents are more likely both to drink bottled water and to have healthy children; they have the stability and wherewithal to offer good food, clothing, shelter, and amenities. Families that own cappuccino makers are more likely to have healthy babies for the same reason. Making a practice of questioning correlations when reading about "links" between this practice and that condition is good statistical hygiene.

In other cases, the mystery of statistics is not a result of psychological blinders or mathematical esoterica, but of a lack of information about exactly what they mean and how they were obtained. The above-mentioned 10 percent condom-failure rate, originally cited in a Planned Parenthood study, is one example. It appears that it resulted from asking couples what their primary method of birth control was and whether it had ever failed them. Approximately one out of ten condom-using couples answered yes, and a statistic was born, even though there seem to be no other statistics to back up the figure.

If the issue is condom leakage, these rates are exceedingly low. (Based on its own and other investigations, *Consumer Reports* concluded: "In principle, latex condoms can be close to 100 percent effective.") If the concern is contraception, the data depend significantly on age, race, and marital status, categories that are surely independent of condom-failure rates. Likewise, if the question is the prevention of sexually transmitted diseases, the numbers again depend on how carefully the condoms are used, but such figures are, except for voyeurs perhaps, difficult to estimate. There is considerable circumstantial evidence, however: Prostitutes in Nevada whose clients always use condoms, for example, contract almost no sexually transmitted diseases.

Failure to put statistics, even accurate ones, into context makes it almost impossible to evaluate personal risk with a clear eye. For example, we often hear the claim that 1 in 8 women will develop breast cancer. This figure is correct, but it is misleading for several reasons. First, it is a lifetime incidence risk, not a mortality risk, which is 1 in 28. Second, the incidence rate for breast cancer, like that for most cancers, rises with age; the risk of a woman's having developed breast cancer by the age of fifty is 1 in 50, but by the age of eighty-five it is 1 in 9. According to a 1993 report by the National Cancer Institute, the typical forty-year-old woman has about a 1.58 percent chance of developing the disease before she reaches fifty and a 3.91 percent chance of developing it before sixty. The typical twenty-year-old, by contrast, has a 0.04 percent chance of developing the disease before age thirty and a 0.47 percent chance of developing it before forty.

The lifetime risk from breast cancer has risen in the last twenty years, but two factors must be remembered. Increases in screening have led to the early discovery of more cancers, and women are dying less frequently from other causes and hence living to ages where the incidence risk is higher. Interestingly, *any* health practice that correlates positively with longevity will likely correlate positively with cancer incidence as well.

This fact brings to mind the true, but potentially deceptive, statistic that heart disease and cancer are the two leading killers of Americans. Of the approximately two million Americans who die each year, for example, almost half die from cardiovascular diseases and about one-fourth from malignancies of various sorts. But accidental deaths—falls, car accidents, drownings, poisonings, fires, gun mishaps, and the like—result in slightly more *years* of potential life (subtracted from the conventionally chosen age of sixty-five), according to the Centers for Disease Control. The average age of accident victims is much lower than that of victims of cancer and heart disease. Along this dimension, AIDS and murder loom ominously as well. For all such circumstances, the number of dead is smaller, but the number of years of life lost is greater.

One final note: Implausibly precise statistics, like the ones I fooled with at the beginning of this segment, are often bogus. Consider a precise number that is well known to generations of parents and doctors: the normal human body temperature of 98.6° Fahrenheit. Recent investigations involving millions of measurements have revealed that this number is wrong; normal human body temperature is actually 98.2° Fahrenheit. The fault, however, lies not with Dr. Wunderlich's original measurements—they were averaged and sensibly rounded to the nearest degree: 37° Celsius. When this temperature was converted to Fahrenheit, however, the rounding was forgotten, and 98.6 was taken to be accurate to the nearest tenth of a degree. Had the original interval between 36.5° Celsius and 37.5° Celsius been translated, the equivalent Fahrenheit temperatures would have ranged from 97.7° to 99.5°. Apparently, dyscalculia can even cause fevers.

Asbestos Removal Closes NYC Schools

Contaminated Mountains Out of Mole Spills

A recent study concluded that 45 percent of science stories in the news deal with medicine and health, another 15 percent with the environment, and 8 percent with the social sciences. Since up to $^2/_3$ of all science articles concern topics linked directly or indirectly with dangers and hazards, better ways must be found to communicate degrees of risk to the general public. (Note the sneaky phrase "*up to* $^2/_3$.")

One particular sort of risk that receives inordinate attention involves contamination: benzene in Perrier water, pesticide residues on vegetables, Alar on apples, asbestos in schools, and chemicals in the soil, water, and air. Although some of the countless contamination risks we hear about may warrant action and justify fear, most are sensationalizing what are essentially trivial hazards, far less worrisome than more mundane risks. (A parody of the misinformed American family springs to mind: Mother, grotesquely obese, sits in the front seat of the family car munching on potato chips. Father alternately sips from a can of beer and puffs on his cigarette while maintaining two fingers on the steering column, and Junior stands between them playing with the rearview mirror. All the while they bemoan the poisons, chemicals, and residues around them.)

To be fair, it often takes time to weed out the noxiously chemical from the innocuously chimerical. But both make some people apoplectically, mindlessly irate, and accounts of their actions make up the bulk of the news reports on the topic. Although difficult, it is important to make distinctions: If everything is risky, nothing is. And, as I mentioned earlier, if unlimited amounts of money are spent on inconsequential hazards, little will be available for significant ones.

A little calculation illustrates how small an amount of contaminant is required to give the impression of a serious risk. Assume the earth's oceans contained pristinely pure water and that some environmental demon were to spill into them a pint of some awful chemical—say, Li_2O for the sake of fantasy—and then systematically churn them up, so that the chemical was evenly distributed throughout. (A liquid pint is a bit more than the volume of a typical can of soda.) A few years later an inspector from an environmental agency removes a pint of water from an ocean somewhere and indignantly announces that there are X molecules of Li_2O in this pint of formerly pure water. What would be your guess of the approximate value of X?

Let me sketch for you how to use arithmetic, a smidgen of geometry, and a smattering of chemistry to come up with a very rough order-of-magnitude estimate of this number. (Skip this and the next two paragraphs if you abhor this kind of stuff.) Note first that the surface area of the earth is approximately 2×10^8 square miles. (The radius, r, of the earth is about 4,000 miles, and the surface area of a sphere is $4\pi r^2$.) Knowing that 75 percent of the earth's surface is covered with water at an average depth of about 2 miles, we determine that the volume of water in the world's oceans in cubic miles is 3×10^8. Multiplying this figure by $5,280^3$, the number of cubic feet in a cubic mile, we find that the volume of the water in the world's oceans is, in cubic feet, about 4.4×10^{19}. Since there are about .017 cubic feet in a pint, the volume of the ocean is approximately 2.6×10^{21} pints.

Continuing, note that there are about 29 cubic inches per pint and roughly .06 cubic inches in 1 cubic centimeter; thus there are approximately (29/.06 =) 480 cubic centimeters in a pint of water or, equivalently, 480 grams of water, or, using the fact that a mole of water weighs about 18 grams, about 25 moles of water in a pint.

Each mole of water contains Avogadro's number (6×10^{23}) of molecules, so a pint of water contains 1.5×10^{25} molecules of water. (There are more direct routes to this number, whose size explains why it is so easy to make a mountain out of a mole spill.)

So a pint of the now polluted ocean contains how many molecules of Li_2O? The fraction of the ocean's volume that is Li_2O is $^1/2.6 \times 10^{21}$). And this is also the fraction of the chemical in a pint of ocean. Since a pint contains about 1.5×10^{25} molecules, we multiply these two numbers and see that almost 6,000 molecules of the vile Li_2O reside in every pint of the world's oceans.

That pint of Li_2O (a volume slightly bigger than that of a soda can, remember) dropped into the pure oceans of the world and spread about uniformly resulted in almost 6,000 molecules of the stuff appearing in every pint we retrieved. The point of this tiny orgy of calculation and dimensional analysis is that it doesn't take much to come up with a frightening headline. One part out of 2.6×10^{21} probably doesn't sound like much even to an alarmist, but 6,000 molecules per pint would almost certainly rouse anxiety among many.

Moving from aqueous to terrestrial dimensional analysis, I remember when very young reading advertisements on the backs of cereal boxes offering land for sale in Colorado. The seemingly dirt-cheap 1950s price was only 25 cents per square inch. Despite my fear of cows and horses, I envisioned my family selling our house and, with the proceeds, buying a cattle ranch so vast it faded into the sunset. What would an acre of this undoubtedly desolate land cost? A square mile?*

In general, almost any mathematically expressed scientific fact can be transformed into a consumer caveat (or lure) that will terrify (or attract) people. This is especially so in a population one-third of whose members believe that humans and dinosaurs lived at the same time, as a recent Louis Harris study has found. *The Journal of Irreproducible Results*, a satirical science periodical, once ran an article illustrating this. Warning: The mass in this product is equivalent in

---

*Since an acre contains 43,560 square feet, one acre would cost $43,560 \times 12^2 \times \$.25 = \$1,568,160$; and since there are 640 acres per square mile, this wonderful Colorado real estate would go for just over a *billion* dollars per square mile (at late 1950s prices).

energy to Q million tons of TNT. Warning: This product attracts every other item in the universe with a force equal to the product of their masses divided by the square of the distance between them. Or, a warning in the same spirit of the example above: Apple seeds contain easily detectable amounts of cyanide, a chemical known to be toxic to humans.

Super Collider a Waste of Money
Science Journalism and Advocacy

There are movie critics, book critics, fashion critics, and restaurant critics, but there are few science critics. The educational specialization required usually makes science writers, or scientists outside their narrow band of expertise, diffident about criticizing any scientific project. They can be too easily rebuked for discussing matters beyond their understanding, and in their natural reluctance to incur the resentment of sources and/or colleagues, they sometimes lose all critical distance. (In a 1994 *New York Times* op-ed on congressional failure to support the Texas Super Collider, the science writer Dick Teresi characterized a physicist who had given up the subject to become a science writer as analogous to Donald Trump's giving up finance to become a bellhop.)

Occasionally the opposite occurs and a scientist or science writer who is not diffident enough will propound a half-baked theory radically at odds with that of the vast majority of scientists in the field. Such was the case with Marilyn vos Savant, the author of a widely read and entertaining column in the Sunday supplement *Parade* magazine. Both in her column and in a book of hers that I reviewed for the *New York Times,* she emphatically declared that the Princeton mathematician Andrew Wiles's recently announced proof of Fermat's last theorem was fatally flawed because it ran afoul of some theorems in non-Euclidean geometry. Alas, Ms. Savant badly misinterpreted these theorems, which were irrelevant in any case. Despite the protests and derision of many mathematicians, however, she never retracted her thesis.

If Wiles's proof doesn't pan out, her readers will likely mistake her confusion for prescience, another instance of the embattled layperson besting the arrogant experts. This is because of her treatment in her column several years ago of the so-called Monty Hall problem. The rules of the game show on which this problem was based were vague, but she was essentially correct in her analysis of it. A few of the mathematicians who wrote her were neither correct nor polite, and the matter received a good deal of attention.

Even when the science writer's credentials or stance are not an issue, it's impossible for him or her to assume that the audience shares the basic background knowledge that food critics, for example, can count on.* The problem is not just knowledge of the discipline in question but basic understanding of issues in logic and the philosophy of science. What's the difference between an empirical statement and an a priori one, between scientific induction and deduction? What about the difference between scientific induction and mathematical induction? Does some implication go both ways or is its converse false? Is such and such a claim falsifiable? We need look no farther than the perennial appeal of pseudoscientific garbage, now being presented in increasingly mainstream forums (such as the preposterous NBC program† maintaining that homeopathy cured the bubonic plague) for confirmation of this problem. In addition, there is an unfounded belief that impersonal science is inconsistent with advocacy, or opinion of any sort.

---

*This is probably true even when the audience consists of members of Congress. The science fiction story "A New Golden Age" by the mathematician Rudy Rucker is relevant here. It tells of a group of mathematicians some time in the future who are eager to obtain funding for pure mathematics research from a philistine, computer-crazed Europarliament. They invent a machine that can convert theorems into musical interludes that will painlessly convey mathematical understanding through sound. In a crash program to meet the grants deadline, they make two musical tapes, one of elementary Euclidean geometry, the other of advanced set theory. To humor the husband of an influential government official, they also make a tape of his work, a nonsensical piece of pseudomathematics. They play the tapes for the appropriations committee, which finds the geometry and set theory tapes boring but responds enthusiastically to the ludicrous one. Impressed with the musical rendition of all its flashy symbols and nonsequiturs, the committee awards significant funds for pure mathematical research. The moral is familiar to writers of grant proposals.

†The NBC show *Ancient Prophecies* is another manifestation of this. One of the funnier parts of the show occurs at the end when the voice-over encourages viewers to tune in next time for an all-new episode of *Ancient Prophecies*.

Despite these obstacles, editorials on science or public face-offs between opposing scientists and/or technical journalists should, I think, be encouraged. One possible topic might be the relative value of molecular work on cancer causation and epidemiological studies of its incidence. The latter research, while not fundamental, is cheap and promises to save lives if it can pinpoint widespread behaviors or localized sites associated with cancer. A small diversion of funds from basic molecular research might pay off handsomely. A couple of people who knew the areas well could easily write an engaging point/counterpoint piece that would also be informative. Similar public discussion of the relative merits of super accelerators, the human genome project, or travel to Mars might also be appropriate.

Articles containing such informed advocacy might be useful in stimulating debate about and possibly even interest in real science—and they needn't appear only in the science section of the paper. Despite the heartening fact that these sections are popular with readers, their number is shrinking; aside from computer manufacturers and software houses, advertisers don't like them.

Harvard Psychiatrist Believes Patients Abducted by Aliens

Mathematically Creating One's Own Pseudoscience

In addition to the occasional piece of informed advocacy, science reporting ought every so often to gently debunk pseudoscientific extravagances. The enduring obsession with guardian angels and statues that bleed or cry are cases in point. So is the writing of the Harvard psychiatrist John Mack, whose recent book chronicles (and more or less endorses the authenticity of) the experiences of patients of his who assert that they've been abducted by aliens in UFOs. It received wide notice, and both the *Washington Post* and the *New York Times* ran long profiles on him, which, while not exactly credulous, were not exactly incredulous either.

One might have guessed that such a radical claim would have galvanized scores of reporters. *Time* magazine's James Willwerth had no trouble locating Donna Bassett, one of Dr. Mack's patients. Ms. Bassett had insinuated herself into his program and, pretending to be an abductee, concocted wild stories that made their way into Mack's book. And James Glieck in *The New Republic* wrote a scathing review of the book and what he termed its weaselly dodges and equivocations. Of course, *The Skeptical Inquirer*, whose reason for existence is the critical examination of such claims, also printed an article on the book. (As a fellow of the Committee for the Scientific Investigation of Claims of the Paranormal, which publishes *The Skep-

tical Inquirer, I may be biased, but in my estimation the publication deserves a Pulitzer Prize for its work on these issues over the years.)

Sometimes mathematics is also helpful in uncloaking pseudoscientific claims and explaining their appeal. I wrote earlier about the plethora of coincidences that become apparent when one skims a paper, flicks through a magazine, and channel-surfs through cable television (not to mention simply living one's life). These remarkable relationships between totally dissimilar items frequently seem to have the air of scientific hypotheses: sunspots and the stock market, hemlines and presidential elections, Super Bowl outcomes and the economy. Very often there is some personal connection or some element of self-reference involved in these relationships. (This morning I prefixed the file names associated with this book with MRN for "Mathematician Reads Newspaper," and twenty minutes later I learned that former President Nixon had died. The undoubtedly cosmic connection between these two events is that Nixon's initials, RMN, are a permutation of MRN.) The sheer number of such possible links and associations should convince one that almost all are merely coincidences.

Rather than rehash the skeptical arguments, let me present a mathematical recipe anyone may use to develop his or her very own personal pseudoscience. The method comes from the Dutch physicist Cornelis de Jager, who used it to advance a theory about the metaphysical properties of Dutch bicycles.

The recipe: Take any four numbers associated with you (height, weight, birthday, social security number, whatever you like) and label them X, Y, Z, and W. Now consider the expression $X^a Y^b Z^c W^d$, where the exponents a, b, c, and d range over the values 0, 1, 2, 3, 4, 5, $1/2$, $1/3$, π, or the negatives of these numbers. (For any number N, $N^{1/2}$ and $N^{1/3}$ equal the square root and cube root of N, respectively, and N to a negative exponent, say, N^{-2}, is equal to one over N to the corresponding positive exponent, $1/N^2$.) Since each of the four exponents may be any one of these seventeen numbers, the number of possible choices of a, b, c, and d is, by the multiplication principle, 83,521 ($17 \times 17 \times 17 \times 17$). There are thus this many values for the expression $X^a Y^b Z^c W^d$.

Among all these values, there will likely be several that equal, to two or three decimal places, universal constants such as the speed of light, the gravitational constant, Planck's constant, the fine structure constant, and so on. (If there are not, the units in which these constants are expressed can be altered.) A computer program can easily be written that can determine which of these universal constants is equal to one of the 83,521 numbers generated from your original four.

Thus you might learn that for your choice of X, Y, Z, and W the number $X^2 Y^{[7/3]} Z^{-3} W^{-1}$ is equal to the sun's distance from the earth. Or you might discover any of a host of other correspondences between your personal numbers and these universal constants–all without having to undergo the rigors of alien abduction. De Jager found that the square of his bike's pedal diameter multiplied by the square root of the product of the diameters of his bell and light was equal to 1,816, the ratio of the mass of a proton to that of an electron. Incidentally, the ratio of the height of the Sears Building in Chicago to the height of the Woolworth Building in New York is the same to four significant digits (1.816 versus 1,816) as the latter ratio. Salacious versions of this game are possible, of course. Compare it also to the coincidental similarities, described in the first section, that can be found to link any two American presidents.

That such accidental linkages and their more mundane, nonnumerical cousins are extremely common is not fully appreciated. As a result hucksters can cash in on people's tendency to ascribe significance to any meaningless coincidence. Consider the foreign con artist whose lure was that he could help students hoping to enter a very competitive national university. The man bragged that he knew the arcane details of the admission process, had contacts with the appropriate officials, and so on. After gathering detailed information from the students, he collected an exorbitant fee, promising to return it if the student was not admitted. Every year he threw their information out, yet every year some of the students got in anyway. Their fees he kept.

Similar stories can be told about miracle medical interventions whose efficacy is only an illusion. Many people simply improve on their own. As in the case of random walks and the stock market, it's quite easy to see patterns, especially if one wants very much to find them. The lesson for all of us is that talking only to people who have a vested interest in some result or linkage can be beguiling (especially if that person is a Harvard professor of psychiatry); gullible journalism is often the result.

Of course, coincidences occasionally point up valuable yet overlooked connections or, vastly less often, defective scientific laws. But, as the philosopher David Hume observed more than 200 years ago, every piece of evidence for a miraculous coincidence–that is, for a contravention of natural law–is also evidence for the proposition that the regularities that the miracle contravened are not really laws of nature after all. Any near-instantaneous transmission of voices across great distances might have been considered a miracle centuries ago. As it has turned out, however, the scientific principles prohibiting or seeming to prohibit such transmission were not, in fact, natural laws.

Not only are coincidences not miraculous but–Freudians, tabloids, and popular sentiment to the contrary–an overwhelming majority of them have no significance whatsoever. Neither, I might add, will the expected numerological fatuities connected to the turn of the millennium in 2000 (2001, for purists). Since 1998 equals 3 times 666, the festivities may begin even sooner.

FDA Caught Between Opposing Protesters

Statistical Tests and Confidence Intervals

Two (more) statistical principles should be a part of every newspaper reader's protection kit. The first deals with the trade-offs inherent in any probabilistic situation. Suppose, for example, I hypothesize that at least 20 percent of the people in a certain region are blond but that, upon observing one thousand people in representative parts of the region, I note only eighty blonds among them. Using probability theory I calculate that, given my assumption, the likelihood of this result is well below 5 percent, a commonly used "level of significance." Therefore I reject my hypothesis that 20 percent of the people in the region are blond.

There are two sorts of errors that can be made in using this or any statistical test; they're called, with all the lyricism for which statisticians are known, Type I and Type II errors. A Type I error occurs when a true hypothesis is rejected, and a Type II error occurs when a false hypothesis is accepted. Thus if I don't realize that most of the blonds in the region stay indoors on sunny days, then in rejecting the true assumption about their constituting at least 20 percent of the population, I would be making a Type I error. On the other hand, if a large number of blonds from several Swedish tour groups were to pass before me and I therefore accepted the false assumption about their proportion of the population, I would be making a Type II error. Even when great care is taken to assure random samples, the possibilities of these two types of error always exist.

The FDA, to revert to the headline that opens this segment, must balance the risk of making a Type I error (not okaying an effective drug) with the risk of making a Type II error (okaying an ineffective or harmful drug). It's not surprising that different groups view these two courses of action differently and will protest different actions.

Succinctly: Any statistical test allows for two types of error—rejecting a true hypothesis and accepting a false one. Depending on the situation, we must evaluate the probabilities of these two kinds of error and act accordingly.

The second tool I want to mention is confidence intervals. More difficult than making statistical estimates is deciding how much confidence we should place in the estimates we make. If the sample is large, we can have more confidence that its characteristics are close to those of the population as a whole. If the distribution of the population is not too dispersed or varied, we can again have more confidence that the sample's characteristics are representative.

It's possible to come up with numerical intervals indicating how likely it is that a characteristic of the population sample reflects the population as a whole. Thus we might say that a 95 percent confidence interval for the percentage of voters favoring proposition X is 45 percent plus or minus 6 percent. This means that we can be 95 percent certain* that the population percentage is within 6 percent of the sample percentage—in this case, between 39 percent and 51 percent of the population favor proposition X. Or we might say that a 90 percent confidence interval for the percentage of consumers preferring on-line service Y is 13 percent plus or minus 8 percent, meaning that we can be 90 percent certain that the population percentage is within 8 percent of the sample percentage—in this case, between 5 percent and 21 percent of consumers prefer service Y. If we increase the size of the sample, we can narrow the interval and increase our confidence that it contains the population percentage (or whatever

---

*Since certainty does not admit degrees, "95 percent certain" is a slight but common abuse of the word. It's reasonable to conclude that being 95 percent certain must entail being 5 percent uncertain, but the latter notion is suspiciously like being a little bit pregnant.

the characteristic or parameter is), but it costs money to increase sample sizes.

Survey or poll results that don't include confidence intervals or margins of error are often misleading. Even when surveys do include such confidence intervals, they don't always make it into the news story. Hedging or uncertainty is rarely newsworthy. If the headline reads that unemployment has declined from 7.3 percent to 7.2 percent and doesn't say that the confidence interval is plus or minus .5 percent, one might get the mistaken impression that something good has happened. Given the sampling error, however, the "decline" may be nonexistent, or there may have even been an increase. If margins of error aren't given, a good rule of thumb is that a random sample* of one thousand or more gives an interval sufficiently narrow for most purposes, while a random sample of one hundred or less gives one too wide for most purposes.

Succinctly: Confidence intervals give us a band within which the true value of the characteristic in question will lie with a certain probability.

One last related point concerns the increasingly common practice of blurring the distinction between scientific polls and informal ones, the latter frequently not much better than an assemblage of similar anecdotes. Too many stories (usually local) are prefaced with something like, "This is not a scientific poll, but a sampling of your friends and neighbors who called our special 900 number..." A more accurate prologue might be, "Here is what a few of the most fervent people in your community think about this issue." Confidence intervals for such "samplings" should be quite broad to accommodate their considerable hot air.

---

*Even a nonrandom "sample" of 250,000,000 can have problems. The U.S. Census Bureau insists on trying to count every single one of us and refuses to supplement its count with standard sampling techniques. The result is a serious underestimate, especially of the urban poor.

Senators Eye Hawaii Health Care Plan

Scaling Up Is So Very Hard to Do

Moving from the local or personal to the national or impersonal is a feat. It's not just a matter of "the same thing, only bigger." Friendly, considerate people may not finish last in public life, but neither are they noticeably more effective than the disagreeable, ruthless ones. Strategies and designs useful in small-scale situations often fail when tried on a broader canvas.

An example is provided by the headline above. Hawaii is a tiny archipelago with a little more than one million people, so any health program that is effective there is not necessarily going to work in a vast country with a population more than 200 times as large. Likewise, the story of a flourishing small company that expands rapidly and becomes an unwieldy large one is familiar. Less well known (except to graph theorists) is the fact that adding new roads to a traffic system can sometimes lead to more, not less, congestion. Problems and surprises arise as we move from the small to the large, from the individual to the societal, from the neighborhood to the world. Social phenomena generally do not scale upward in a regular or proportional manner.

Similar observations hold for natural phenomena. Biodiversity rises with the increased availability of food in a region, but only up to a point. Then it declines. Nothing is more basic than such physical notions as length and weight, but even here the transition from small to big is not simple. If a six-foot man were scaled up to thirty feet, for example, his weight, which, like his volume, varies with the *cube* of his height, would go from 200 pounds to 25,000 pounds (200×5^3). The supporting cross-sectional area of his thighs would vary only with the *square* of his height, so the pressure on them would be crushing and the man would collapse. (This is why heavy land animals like elephants and rhinos have such thick legs.) If a teenage boy who weighs 150 pounds were to grow from 5 feet 4 inches to 5 feet 9 inches, how much weight would he gain (assuming his growth were proportional)?*

Comparable but more difficult obstacles arise when we attempt to extend our understanding of human actions and behavior. There is a need for impersonal principles analogous to the length and weight considerations. The most theoretical such principles involve scientific induction, which clarifies the problems associated with inferring general statements from particular observations. Ethical theories, whether based on fixed rules or utilitarian consequences, are sometimes helpful in expanding one's concerns from the family to society. So are the prisoner's dilemma and the tragedy of the commons, puzzles that elucidate the trade-offs between pursuing one's personal interests exclusively and cooperating with the larger group. A typical example might be a group of fishermen who can voluntarily restrict their own take and thrive for decades or else catch the maximum now and collectively do much worse in five years' time.

Another useful tool is Arrow's theorem and related results, which examine the quandaries that arise in moving from the preference rankings of individuals to a preference ranking for a larger group. Recall the earlier Tsongkerclintkinbro segment. For another illustration of the difficulties that may arise, assume that there are three competing health care proposals before the Senate. Assume also that thirty-three of the senators prefer proposal A to proposal B to proposal C, another thirty-three prefer B to C to A, and the remaining thirty-three prefer C to A to B. (Bob Dole likes none of them.) So far, so good. But a problem arises if we look at the Senate as a whole.

---

*Since his height increases by a factor of 69/64, or 1.078, his weight, which varies as the cube of this number, will increase from 150 to $150(1.078)^3$, or to approximately 188 pounds.

Since a total of sixty-six senators prefers A to B, the Senate as a whole can be said to prefer A to B. And sixty-six senators prefer B to C, so the Senate can be said to prefer B to C. The Senate thus prefers proposal A to proposal B, and proposal B to proposal C. You might conclude that the Senate therefore prefers proposal A to proposal C, but you would be wrong. Sixty-six of the senators prefer C to A, and thus the Senate prefers C to A! Even if we assume that each senator is rational and has defensible reasons for his or her ranking, we can't make the same assumption about the Senate as a whole, which, though it prefers A to B and B to C, prefers C to A.

Recent polls provide an instance of a related mathematical problem, that of preserving coherence when moving from a small set of statements to a larger one. The polls showed that support for health care reform was declining at the same time that support for universal coverage was rising. The general mathematical question of determining when a large set of statements is jointly consistent is a most intractable one.

These and other principles and ideas may be somewhat abstract and unappealing, but there are no alternatives if we're to have any hope of scaling up our understanding without buckling under the mental strain. As news services and cable news take over some of the transcriptional aspects of reporting, there is room in newspapers for more analytic, comprehensive accounts to supplement (but certainly not replace) the parochial fights, intrigues, and drama that make up the bulk of the news.

Breakthrough Forecast by
End of Decade

You Can't Know More Than You Know

Freeman Dyson in *Infinite in All Directions* mentions the Hay Theory of History, which maintains that the invention of hay made possible the development of urban civilization in northern Europe. Hay was not required in the days of the Roman Empire because winter weather was mild enough to allow grass to grow and animals to graze all year round. Northern cities, however, depended on horses and oxen for transportation and remained relatively unimportant until the fortuitous invention of hay by some anonymous farmer in the early Middle Ages. The discovery allowed these animals to be kept in large numbers, leading to the growth and mobility of the human population and, eventually, to a shift in the center of power and civilization from Rome to cities like Paris and London. Comparable tales can be told about the invention of the spinning wheel and the printing press.

These technologies and their consequences were unpredictable, and, like technologies before and since, they effectively foreclosed other alternatives. Consider the typewriter. The familiar "qwerty" layout of its keys is a result of mechanical problems that are no longer germane, but its dominance is assured—as even the 486 machine on which I now write attests. DOS, the underlying operating system of my machine, provides another instance of the early locking-in of a technology (although, happily, it doesn't look like it's going to be around forever).

Similarly, the battle between the VHS and Beta format for videocassettes has been won decisively for no very good reason. And, once upon a time, cars with internal combustion engines were no better than ones with steam engines, but, for chance reasons, the former prevailed. Or imagine a parallel civilization almost identical to ours that first developed nuclear power and only later discovered coal power. Those people might well be shocked to discover that the wastes from the new fossil fuels had to be dumped directly into the atmosphere, and would probably elect to retain their cheaper, cleaner fuel.

Even the "technology" of life itself, although neither a discovery nor an invention, is subject to this same pattern of multiple contingencies hardening into accidental universality. It might quite imaginably have been based on a molecule other than DNA. If there were competitive molecules, however, DNA very early won the biotic lottery and firmly established life's unique qwertyness on this planet. The same can be said about the particular course evolution has followed. Really new paths are unpredictable, and if there are alternative candidates, chance and circumstance are generally what determine which one thrives and develops.

This brings me to the headline that opens this segment, which seems to suggest the paradox of scientific prediction. (It is related to Chaitin's theorem, the complexity horizon, and the ideas of Karl Popper on historicism.) If one knows enough to make a prediction of a discovery with any confidence, then one has essentially already made the discovery; only routine checking remains. Equivalently, if one hasn't made the discovery or, what's more common, doesn't even know that there is some indeterminate discovery to be made, then one is not going to be predicting discoveries with any accuracy.

The mistake historicists make in predicting scientific discoveries reminds me of the joke about the college student who wrote his mother that he had taken a speed-reading course. In the middle of her long letter of reply, she wrote, "Since you've taken that course, you've probably already finished reading this letter." The letter isn't finished until it's finished, and the discovery not made until it's made.

And, as with hay, even after the technology or invention has been discovered, forecasting its development and use can be quite difficult, although not nearly so difficult as forecasting the discovery itself. Guglielmo Marconi, who invented the radio, thought it would be used merely as a substitute when a telephone-wire connection was impossible—at sea, for example. In the late 1940s, IBM was not impressed with the commercial possibilities of computers and thought a dozen or so of them would fully satisfy the demand. Bell Labs was initially reluctant to seek a patent on lasers, seeing no use for them in communication, much less in any of the other fields where they've become indispensable. And who can say whether electronic newspapers will remain a small niche market, become adjuncts or companions to standard papers, or, horrors, eventually replace newsprint altogether?

Such unpredictability is a strong argument, I think, for the funding of many relatively small science projects rather than one or two mammoth ones.

Rodent Population Patterns Difficult to Fathom

Ecology, Chaos, and the News

News stories on ecological issues always seem to be on the verge of making what philosophers used to call a category mistake, describing a situation as if it belonged to one logical type or category when in fact it belongs to another. In the case of ecology, there is a tendency to look primarily for culpability and conflicts of human will rather than at the dynamics of a natural process. (I'm reminded of the comedian who announced that an Iranian terrorist group had claimed responsibility for the 1993 floods in the Midwest.) Often, however, there are no discernible agents, no Whos to write about. Furthermore, the process isn't instantaneous or precisely localized, so the journalistic When and Where seem inappropriate as well. Moreover, the What, Why, and How are generally complex and impersonal, often beyond the complexity horizons of most readers and sometimes even of the ecologists themselves.

An intriguingly simple ecological example studied by the biologist Robert May and the physicist Mitchell Feigenbaum illustrates. Imagine that the population of a certain animal species is given by the so-called logistic formula $X' = RX(1-X)$, where X is the population one year, X' the population the next year, and R a parameter that varies between 0 and 4. For simplicity, we take X and X' to be numbers between 0 and 1, the true population being, say, 1,000,000 times these values. If we assume the population X is now .1 (that is,

100,000) and R = 1.5, we can calculate the population next year, X′, by plugging .1 into the formula, obtaining 1.5(.1)(.9), or .135. To calculate the population for two years from now, we plug .135 into the formula and obtain 1.5(.135)(.865), or approximately .1752.

If you have a calculator handy, you might want to verify (or you might not) that the population for three years from now will be .21676. I'll spare you any more arithmetic, but the same procedure may be used to find the population for any future year. You would find that the population grows rapidly and then stabilizes at .3333. Moreover, you would find that whatever its original value, big or small, the population would still stabilize at .3333. This population is termed the *steady state* population for this value of R (see diagram).

If we perform similar calculations for a smaller R, say, R = 1, we find that whatever its initial value, the population "stabilizes" at 0; it becomes extinct. When R is bigger–2.5, to be specific–we find in carrying out these procedures that whatever its initial value, the population now stabilizes at approximately .6, the steady state for this R.

So what? Well, let's increase R a little more, this time to 3.2. As before, the original population does not matter, but now when we substitute iteratively into the formula X′ = 3.2X(1−X), we find that the population of the species doesn't stabilize at one value, but eventually settles into a repetitive alternation between two values: approximately .5 and .8; that is, one year the population is .5 and the next it's .8. Let's raise our parameter R to 3.5 and see what happens. The initial population is once again immaterial, but this time in the long run we have the population alternating regularly among four different values, approximately .38, .83, .50, and .88, in successive years. If we again increase R slightly, the population settles down to a regular alternation among eight different values. Smaller and smaller increases in R lead to more doublings of the number of values.

Then suddenly at approximately R = 3.57, the number of values grows to infinity and the population of the species varies randomly from year to year. (This is an odd sort of randomness, however, since it results from iterating the formula 3.57X(1−X), and the sequence of populations is thus quite determined by the initial population.) Even

Relationship Between Values of R, Population Sizes, and Chaos

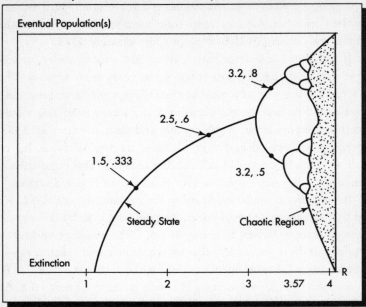

Chaotic region is very intricate. This type of
diagram is derived from the biologist Robert May.

stranger than this chaotic variation in the species' population is the
fact that a slight increase in R again results in a regular alternation of
the species' population from year to year, and a further increase again
leads to chaotic variation. These orderly alternations followed by
random chaos followed again by windows of regularity are critically
dependent on the parameter R, which seems to be a gross measure of
the volatility of the model.

A close-up view of the chaotic region reveals an unexpected intricacy, and, as in the case of the economy discussed earlier, there is a lesson to this intricacy. If a system as trivial as this single nonlinear
equation can demonstrate this sort of chaotic unpredictability, perhaps
we shouldn't be quite as assertive and dogmatic about the predicted
effects of various ecologic policies on the gigantic nonlinear system that
is Planet Earth. The policies' consequences, one would think, are considerably more opaque than are those of the value of R in this model.

Let me reiterate that there are no agents involved in the story, no greedy corporate despoilers or overzealous environmentalists. Any news story that purported to explain the variation in population would either have to attribute it falsely to someone or something comprehensible to most readers or else find some other way to present the account. As in reporting on the economy and other complex systems, there is a tendency to settle upon some stylized, conventional set of stories that focus on turf and clout rather than on the intricacies of the impersonal system in question.

More Dismal Math Scores for U.S. Students

X, Y, and U

Such headlines remind me of the children's riddle about Pete and Repeat.

> FIRST CHILD: Pete and Repeat are walking down the street. Pete falls down. Who is left?
> SECOND CHILD: Repeat.
> FIRST CHILD: Pete and Repeat are walking down the street. Pete falls down. Who is left?
> SECOND CHILD: Repeat.
> FIRST CHILD: Pete and Repeat are walking . . .

Looked at in the right way, it might almost be humorous. Some eminent commission issues a warning that the mathematics (or science) performance of American students is dismal. Then the stylized expressions of alarm are voiced, after which the subject is quickly forgotten until the next, even more dismaying report is announced. Pete and Repeat.

Why should we really care? As a mathematician, I'm often challenged to come up with compelling reasons to study mathematics. If the questioner is serious, I reply that there are three reasons or, more accurately, three broad classes of reasons to study mathematics. Only

A longer version of this piece appeared in the August 7, 1994, issue of the *Washington Post*.

the first and most basic class is practical. It pertains to job skills and the needs of science and technology. The second concerns the understandings that are essential to an informed and effective citizenry. The last class of reasons involves considerations of curiosity, beauty, playfulness, perhaps even transcendence and wisdom.

And what are the costs incurred if we as a society continue in our perversely innumerate ways?

The economic cost of mathematical ignorance is gauged, in part, by people who, though they can perform the basic arithmetical operations, don't know when to do one and when to do another: clerks who are perplexed by discounts and sales taxes, medical personnel who have difficulty reckoning correct dosages, quality control managers who don't understand simple statistical concepts. The supply of mathematically capable individuals is also a factor in the U.S.'s position in many new scientific technologies, among them fuel-efficient engines, precision bearings, optical glasses, industrial instrumentation, laser devices, and electronic consumer products.

As Labor Secretary Robert Reich and others have written, those jobs and job classifications requiring higher mathematics, language, and reasoning abilities are growing much more rapidly than those that do not. (It should be admitted, however, that although their rate of growth is higher, the number of such jobs still doesn't compare with the huge number that require lesser skills.) Nevertheless, enrollment in college-level mathematics is down, and fewer and fewer American students are majoring in math or in the growing number of fields that require it.

The social cost of our mathematical naïveté is harder to measure (although I try to in this book), but gullible citizens are a demagogue's dream. Charlatans yearn for people who can't recognize trade-offs between contrary desiderata; who lack a visceral grasp of the difference between millions of dollars for the National Endowment for the Arts and hundreds of billions of dollars for the savings-and-loan bailout; or who insist on paralyzing regulation of rare and minuscule health risks, whose cumulative expense helps to ensure the incomparably greater health hazard of poverty. As this book shows, almost every political issue—health care, welfare reform, NAFTA, crime—has a quantitative aspect. And, as I mentioned earlier and as the FDA is just now recognizing, the worst drug problem in this country is not crack or cocaine but cigarette smoking, which kills 400,000 Americans annually, the equivalent of three fully loaded jumbo jets crashing each and every day of the year. Or consider Lani Guinier's mathematical suggestions regarding the Voting Rights Act, or the possible economic and ecological implications of chaos theory, or the statistical snares inherent in the interpreting of test results, whether they be for academic achievement or the presence of drugs.

Regarding the third class of reasons, I think it's only fair to say that the "cost" of the philosophical impoverishment resulting from mathematical illiteracy is one that millions of Americans gleefully assume. Still, there is evidence that people respond enthusiastically to mathematical topics as long as they are not labeled as such. People enjoy complicated numerical and mechanical puzzles (Rubik's Cube, for example, the number of whose possible states—4×10^{19}—is greater than the distance in inches that light travels in a century); crossword puzzles and word play (including certain kinds of humor); board games; all sorts of gambling; paradoxes and brain teasers. Part of the enjoyment is traceable to their quasi-mathematical charm. We have an innate attraction to pattern, structure, and symmetry that mathematics and science develop and refine. Certainly Bertrand Russell is not alone in prizing the subject's "cold and austere beauty," and many are excited by Andrew Wiles's likely proof of Fermat's last theorem, even if it is of little earthly use.

With these reasons to study mathematics, why don't American students do better? Enough's been written about our many social problems, so rather than plunging into that dreary story let me end this segment with five misconceptions about mathematics. Cumulatively, they contribute significantly to poor pedagogy in school and needless handicaps on the job.

Probably the most harmful misconception is that mathematics is essentially a matter of computation. Believing this is roughly equivalent to believing that writing essays is the same as typing them. Or, to vary the analogy, imagine the interest in literature that would be engendered if every English class focused exclusively on punctuation. Of course, this is not to say that calculating mentally or with paper and pencil is not important. Nor is it meant to discourage the excellent habit of estimating quantities. It is merely to assert that in mathematics, as in other endeavors, the big picture is seldom presented.*

Another misconception about the subject is that it is strictly hierarchical; first comes arithmetic, followed in lockstep by algebra, calculus, and differential equations, after which arrive abstract algebra, complex analysis, and so on. There is undeniably a cumulative aspect to mathematics, but it is less significant than many think. A third misconception concerns what, for lack of a better term, I'll call storytelling. An effective educational strategy since ancient times, storytelling places a question into context, demonstrates its relation to other questions, and concisely lays out some seminal idea. The rigid view that draws a sharp distinction between formal math and narrative may explain why this plain means is too seldom employed in mathematics education and why the topic of this book will seem strange to many people.

Parental expectations can be effective in thwarting the effects of the next misconception, which is frequently signaled by comments such as "I'm a people person, not a numbers person" or "Math was always my worst subject." Although it is undoubtedly true that some people have considerably more mathematical talent than others, mathematics is not only for the few. There are also disparities in writing ability, but people rarely counsel students to give up on their English courses because they're not going to make it as novelists.

The last misconception is the romantic belief that a concern with mathematics is somehow numbing, making one unresponsive to,

---

*Although details are very often critically important, an inability to stand back and "chunk" facts leads to a myopic favoring of minutiae over ideas in many contexts. As I've mentioned, computation is valued above conceptual understanding in mathematics; in politics, smart tactics bring greater rewards than wise policies; technical hocus-pocus in the stock market attracts more attention than does analysis of fundamentals; for those with a religious temperament, rules, rites, and rituals obscure wonder, awe, and mystery; in sex, lust and fetishism are mistaken for love. I grant that the first element in each of these oppositions does sometimes rightfully take precedence over the second, but generally too little stress is placed on the second. It's much easier to put the jigsaw puzzle together after you've seen the whole picture (assuming there's one to see).

say, stone farmhouses in the late afternoon sun. Asking How Much, How Many, or How Likely is thought to make one a member of Napoleon's despised "nation of shopkeepers." Sentiments such as these are as potent as they are baseless.

I'll forgo discussing the curricular and pedagogical suggestions that follow from this discussion, except to note that the newspaper is an undervalued source of examples and ideas for mathematics classes at various levels. Once an idea or notion is grounded in some real-life situation, it can later be generalized, idealized, and aestheticized.

Pete and Repeat . . . Let's not.

Section 5

FOOD, BOOK REVIEWS, SPORTS, OBITUARIES

It is frivolous stuff, and how rare, how precious is frivolity!
—E. M. FORSTER

Contemporary newspapers are full of special features, ranging from the fashion and travel pages to sports news and advice columns. There are also lucky numbers for the lottery, numerological and biorhythmic flummery, weather forecasts that sometimes defy interpretation, personals ads with their own arcane zoology (SDJMNS), and lists, surveys, and fillers of all sorts. It's easy to be snide about (some of) them, but I think they help make the paper excitingly alive and multifaceted. Besides, rigid distinctions between the deep and the shallow are generally themselves quite superficial; there are a few serious issues underlying the segments that follow and a few superficial issues underlying the preceding ones. A good illustration of the inseparability of the serious and the silly is the fact that Gary Trudeau's cartoons appear not only on the back comics pages of many papers but also on the op-ed page of the *New York Times*.

What follows are very short responses to various short features, some dribs on squibs, if you will (or even if you won't).

761 Calories, 428 Mgs. Sodium, 22.6 Grams of Fat per Serving

Meaningless Precision

I seldom read the food section of newspapers. The reviews of restaurants frequently seem pretentious or overly cute, and I'm not interested in the recipes. I do have one mathematical observation regarding the latter, however. I will sometimes read a recipe that calls for about a cup of this, a few tablespoons of that, a smidgen of something, four or five slices of something else, a couple of medium-sized other things, and seasoning to taste. This vagueness is not objectionable, but the arithmetic that comes from it is. In italic print at the end, it's affirmed that the recipe yields, say, four servings with 761 calories, 428 milligrams of sodium, and 22.6 grams of fat per serving. These numbers are much too precise, given what went into computing them. The 1 in the 761 is completely meaningless, and the 6 is almost so. The 7 is the only significant number. Saying that each serving is 700 to 800 calories would make more sense, and an estimate of 600 to 900 calories would be even better. An added benefit to having a reasonable range of numbers rather than a figure of unwarranted precision is that one isn't as tempted to swell the recipe and pretend that one is consuming only 761 calories.

I've also noted quite significant variations in the size of a frozen yogurt cone or in the amount of cereal in a box. I suspect that what's true for individual recipes is true for packaged and frozen foods as well, although perhaps to a lesser degree due to more uniform manufacturing standards. In any case, the problem of baselessly exact numbers transcends food and recipes. A neighbor, who knows I'm a mathematician, tells me proudly that he gets 32.15 miles per gallon, leaving unclear whether his pride is in his car or his arithmetic prowess. My daughter's teacher awards her a grade of 93.5 on an essay, leaving visible the 94 that he crossed out.

If the number at issue is a sum or product, or is otherwise mathematically dependent on several other numbers, only one of them need be imprecise for this imprecision to infect the given number. The joke about the museum guard who told visitors that the dinosaur on exhibit was 90,000,006 years old is a good illustration. Upon questioning, the guard explained that he was told the dinosaur was 90,000,000 years old when he was hired, six years before.

Top Designs for the Busy Working Woman

Fashion, Unpredictability, and Toast

The fashion pages have always puzzled me. In my smugly ignorant view, the articles appear to be so full of fluff and nonsense as to make the astrology columns seem insightful by comparison. The models are beautiful, of course, but their clothes often look unwearable by anyone other than a model. This is especially true when their price tags match those of good used cars. Always risible are the claims of the "top designers" that these glitzy, outlandish concoctions are for the busy working woman.

Still, even I realize there is such a thing as fashion, and many people seem to know what is in or out of style. But what determines what is fashionable today and, more interestingly, what will be so tomorrow? In the case of the weather, the economy, or an ecosystem, we generally have a good understanding of what most of the important variables are, and these systems (at least the weather and probably the others) are nevertheless subject at times to deterministic chaos and unpredictability. In the case of styles and fashions, however, even the most important variables are unknown (at least to me). Moreover, their jittery, self-referential responsiveness to countless influences is so convoluted that any attempt at long-term anticipation is hopeless. After-the-fact accounts of the development of a style are possible, but not forecasts of the future. It's an appealing thought that the fashion pages of a few years hence are beyond not only my complexity horizon, but that of top designers too.

Having just mentioned chaos and, before that, recipes, I think the time is right to describe an insight I had recently. It is to Proust's epiphany what soggy toast is to petite madeleines, but let me expatiate upon it a bit. My favorite breakfast is toast, jam, juice, milk, and Diet Coke, which I consume distractedly as I read the paper. While thus occupied one recent morning, I recalled the following (variant of a) technique devised by the mathematician Steve Smale to illustrate the evolution of chaos. Imagine the toast moistened, folded, and compressed into a cubical piece of white dough through the middle of which runs a thin layer of raspberry jam. Now stretch and squeeze this sandwich to twice its length, then fold it smoothly back upon itself to reform the cube. The jam layer is now shaped like a horseshoe (see diagram).

Repeat this stretching, squeezing, and folding a large number of times and you'll notice that the jam (I'm idealizing here) is soon spread throughout the dough in a convoluted pattern, similar to a filigreed pastry. Points in the jam that were close are now distant;

Toast, Jam, and Chaos

Points in the jam that were close are now distant;
other points that were distant are now close.

other points that were distant are now close. The same is true for points in the dough. Smale used this "horseshoe" procedure to clarify the advent of chaos in a dynamic system. Furthermore, it has been argued that all chaos (and the consequent ticklish unpredictability it gives rise to) results from such stretching, squeezing, and folding in a suitable logical space.

All right, here's the insight (really a somewhat strained analogy) I promised. Reading a newspaper, among other activities, is an efficient means of doing to our minds what the stretching, squeezing, and folding do to the raspberry jam. The stretching and squeezing correspond to our envisioning of the distant events, different people, and unusual situations that are tersely reported in the paper, and the folding corresponds to what we do if we try to make what we've read a part of our own lives. Every day, our mental landscape is stretched, squeezed, and, if we allow it, folded back upon itself, and the effect on us is similar to that on the raspberry jam. Ideas, associations, and beliefs that were close become distant and vice versa. People keenly attuned to the world and what they read about it are also much harder to predict, I suspect, than those whose range and purview are more limited.

There is little cash value to this metaphor. In this vague form it doesn't make any scientific claim and, in fact, seems almost unfalsifiable. Still, it is suggestive and seems to be consistent with the idea that we ourselves are nonlinear dynamic systems subject to the same chaos and unpredictability as the weather. It does seem, for example, that dark moods sometimes come over us in the way bad storms bring a sudden end to picnics.

Of course, we're not that unfathomable, so there must be countervailing statistical considerations that make for predictability and stability. These are connected to my theory of waffles, so I'll skip them for now.

Agassi Wins Again

Scoring and Amplifying Differences

The sports pages duly, daily, and too often dully record the base-ball, football, basketball, and hockey scores. (There is, in fact, soft-ware that can create very routine sports stories if given the basic results and a few highlights of the game in question.) An obvious fact that's always fascinated me about this unending sequence of contests is that the best teams manage to lose and the worst ones manage to win regularly. Interestingly, however, this is less often the case in individual sports, where the scoring ensures that unless players are reasonably close in ability, the better one will almost always win.

For example, if I were to practice tennis diligently for 109 hours a day, I can imagine winning 40 percent of my points against Andre Agassi. Even given the conditions of this fantasy, however, I would get 4 points before Agassi does just 29 percent of the time. Since to win a game you must obtain at least 4 points and be at least 2 points ahead of your opponent, it turns out that I will win only 26 percent of my games against him. (The scoring is archaic: love, 15, 30, 40, game, rather than 0, 1, 2, 3, 4.) To win a set, you must win at least 6 games and be 2 or more games ahead of your opponent. Thus, even though I win 40 percent of the points and even if I manage to win a game or two or three, my chances of winning a set are less than 5 percent. Furthermore, were I to win a set, I would still have to win 3 out of 5 sets to win a match, and for this my chances shrink to less than one-twentieth of 1 percent. These numbers may be derived using the so-called negative binomial distribution (the details may be found in Ian Stewart's *Game, Set, and Math*).

This is, of course, an idealized model of tennis. One glaring oversimplification results from ignoring the fact that you are more likely to win a point when you are the one serving. Nevertheless, it's undeniable that the rules of the game (and of many other endeavors that aren't normally classified as games) greatly amplify the differences between the participants. Compare this with the discussion of the extremes of the normal distribution in an earlier segment.

Another way to see the amplification effect of repeated play is to imagine flipping a biased coin against an opponent. Assume that the probability of your winning the coin flip, thereby obtaining 1 point, is 25 percent. Now if the person who first attains 10 points is the winner of the game, your chances of victory will be less than 1 percent. If you're playing against an opponent with unlimited wealth, such as a casino, you're sure to lose all your money eventually at this game. (In fact, this probability problem is referred to as the "gambler's ruin.") Your best bet would result if the victor were determined by a single flip of the coin, in which case your chances would be 25 percent. This is the mathematical version of the old proverb that says it's better to talk little and let people merely suspect your ignorance than to talk incessantly and let them know for sure.

To get a flavor for the computations involved, calculate the probability of your winning 2 points before your opponent does if the probability of your winning any given point is just 25 percent. What is the probability of your winning 3 points before your opponent does?*

---

*You will win 2 points before your opponent does if you win two in a row, symbolized by WW, or if you win one, lose one, then win one, symbolized by WLW, or if you lose one, then win two in a row, symbolized by LWW. By the multiplication principle for probability mentioned in section 1, the probability of each of these is $(.25)(.25)$, $(.25)(.75)(.25)$, and $(.75)(.25)(.25)$, respectively. If we sum these three products, we get approximately .156, or $^5/32$ if you use fractions instead of decimals. There are more ways in which you can win 3 points before your opponent does. Symbolized as above, these are: WWW, LWWW, WLWW, WWLW, LLWWW, LWLWW, LWWLW, WLLWW, WLWLW, and WWLLW, and their probabilities sum to approximately .104. When the numbers are bigger, we must use combinatorial techniques to count the various possibilities.

New Survey Reveals Changing Attitudes

Societal Gas Laws

Looking at the daily statistical snapshots feature in the corner of *USA Today*, I wonder at the extent to which polling and measuring have become a national pastime. This mindset that now seems so familiar can be traced to the nineteenth-century Belgian academic Adolphe Quetelet, who wrote:

> Thus we pass from one year to another with the sad perspective of seeing the same crimes reproduced in the same order and calling down the same punishments in the same proportions. Sad condition of humanity! The part of prisons, of irons, and of the scaffold seem fixed for it as much as the revenue of the state. We might enumerate in advance how many individuals will stain their hands in the blood of their fellows, how many will be forgers, how many will be poisoners, almost as much as we can enumerate in advance the births and deaths that should occur.

The impersonal perspective that statistics and polling induce serves as a counterweight to our fascination with individuals, extremes, and anomalies. Although the particulars of a story are frequently all-important, an inability to stand back and average opinions and observations leads to a myopic focus on minutiae, as I have stressed earlier. One of the primary mathematical tools justifying this standing back and averaging is the so-called Central Limit theorem. It states that the average (or sum) of a *large* bunch of measurements follows a normal bell-shaped curve even if the individual measurements themselves do not.

Imagine a factory that produces, say, diet food. Let's suppose it is run by a sadistic nutritionist who ensures that only 60 percent of the packages contain the advertised 500 calories, that 30 percent of the packages contain roughly 2,000 calories, and that 10 percent contain approximately 5,000 calories. The distribution of the calorie content of these packages is clearly not described by a normal bell-shaped curve, but rather by a curve consisting of three spikes, one at 500 calories and two smaller ones at 2,000 calories and 5,000 calories.

Assume that these packages come off the assembly line in random order and are packed in boxes of thirty-six. If a quality control engineer decides to determine the average calorie content of all the packages in a box, he'll find it to be about 1,400 or so–say, 1,419. If he determines the average calorie content of the packages in another box of thirty-six, he'll again find the average content to be about 1,400 or so–perhaps 1,386. In fact, if he examines many such boxes, he will find the average of the calorie contents to be very close to 1,400. More surprisingly, the distribution of these averages will be approximately normal (bell-shaped), with the right percentage of boxes having averages between 1,350 and 1,450, between 1,300 and 1,500, and so on.

The Central Limit theorem states that under a wide variety of circumstances, averages (or sums) of even non-normally distributed quantities will nevertheless themselves have a normal distribution. Other quantities that tend to follow a normal distribution might include age-specific heights and weights, water consumption in a city for any given day, the number of raisins in cereal boxes, widths of machined parts, IQ's (whatever justification they have or lack, whatever it is they measure), the number of admissions to a large hospital on any given day, distances of darts from a bull's-eye, leaf sizes, and the amount of soda dispensed by a vending machine. All these quantities can be thought of as the average of many factors (physical, genetic, or social) and thus turn out to be normally distributed.

There are other ways in which a kind of average order can arise out of disorder. In the area of physics known as statistical mechanics, for example, scientists don't try to trace the trajectories of individual gas molecules in a container, but rather examine various statistical averages associated with the molecules. Here too the averages are quite stable and explain macroscopic properties, such as the temperature in the container. Likewise, rather than saying where each item in my son's room is located, I can instead report that his room is a complete mess. Describing it as such imparts some useful information. For example, don't look for large empty areas or neatly stacked books, and don't be shocked to find scraps of food scattered about. Mutual funds are another example: They're attractive because, properly balanced, they fluctuate much less than the individual stocks they contain and hence are more amenable to forecast than are the individual stocks.

Given any collection of objects and relations among them, there is always some order or pattern that may be ascribed to the system, even if it's only a statement of the objects' randomness. Randomness on one level of analysis constitutes a kind of order on a higher level. Random movements of individual molecules give rise to the ideal gas laws at the macroscopic level. Haphazard arrangements of belongings give rise to the adolescent mess laws. Properly selected funds give rise to portfolio theory.

As Quetelet first observed, a similar idea applies to sociology, sports, sex,* political science, and economics, which may be thought of as a kind of social statistical mechanics. Under this understanding, the old chestnut about polls taking the temperature of the public is quite apt. There are, of course, innumerable abuses, countless possible misinterpretations, and depressingly many biased studies, but, done right, the process works; it yields knowledge.

---

*For individuals, sex is a private event whose frequency varies radically with circumstances: age, relationships, job stress, and so on. Looked at from a planetary perspective, however, the number of incidents of sexual intercourse initiated per hour probably changes very little from one hour to the next, day in and day out, year after year. I would estimate the rate to be about 20 million per hour, with regular dips and rises due to differences in population densities in various time zones. As Stanislaw Lem mentions in *One Human Minute*, what's important is not the number, which is only a guess, but the amazing unrelentingness of the pattern, whatever it might be.

Near-Perfect Game for Roger Clemens

How Many Runs in the Long Run

Sports records provide a superb illustration of the laws of probability and averages. There is good evidence, for example, that the notion of a "hot hand"–a long streak of consecutive successful shots in basketball or of hitting safely in consecutive games in baseball–can be adequately explained by chance alone. (Recall the segment on coins and the market in section 2.) The number, frequency, and duration of such streaks are about what one should expect, given the skills of the players. That is, if a player makes 60 percent of his shots, he'll have approximately as many streaks, and they'll last as long, as would a (biased) coin that lands heads 60 percent of the time. (Marginally relevant is the story I related in the introduction of the three duck-hunting statisticians who used their knowledge of averages to bag a duck.)

A good baseball example involves the pitching of perfect games—no hits, no walks, and no errors. Roughly 30 percent of the time, a batter will get on base somehow. Hence the probability of a pitcher getting a batter out ("retiring" him) is 70 percent–.7, in decimals. The probability of the pitcher's retiring two batters in succession is $.7 \times .7$, or .49 (just as the probability of a flipped coin's landing heads twice in succession is $.5 \times .5$, or .25). The probability that the pitcher will retire twenty-seven batters in succession (three batters per inning for nine innings) and thus pitch a perfect game is $.7^{27}$, or about 1 chance in 15,000. So what?

Let's assume that major-league teams have averaged about 3,500 games per season for decades (counting each game twice, since both pitchers have a chance to pitch a perfect game, and taking into account that until relatively recently there were only sixteen teams playing a shorter season). Hence a perfect game should occur approximately once every 4.3 years (15,000 divided by 3,500), or 9.3 times in forty years. In this case the estimate is uncannily accurate, there having been nine perfect games pitched over the last forty years, during which time there were approximately 140,000 chances to do so. Presumably, if forty years of baseball were played over and over again under roughly the same circumstances, the number of perfect games within any forty-year period would sometimes be seven or eight, sometimes ten, eleven, or twelve, but would average around nine.

Averages aren't as straightforward in the more nebulous realm of baseball business, of course. For a topical instance, consider the baseball owners' constant refrain during the 1994 strike that the average player's salary is $1.2 million. This is true enough, but the median salary is $500,000; half the players earn less than that, half more.

Having mentioned this staple of statistics courses, I end this segment on averages with two slightly tricky exercises on means, medians, and modes. First, consider the salaries of five players: $5,000,000, $4,500,000, $900,000, $800,000, and $800,000. The mean salary (average) of $2,400,000 is greater than the median salary (middle number) of $900,000, which in turn is greater than the mode salary (most common value) of $800,000. Is it possible to find a set of five numbers whose median is bigger than its mode, which is in turn bigger than its mean?*

The second exercise concerns the financier who invests $100,000 in a volatile team franchise that each year, with equal

---

*For larger sets of numbers, any ordering of the three statistics—mean, median, or mode—is possible. But if there are only five numbers and the median is larger than the mode, then the two numbers less than the median must be the same, and this forces the mean to be bigger than the mode. Thus the answer is no.

probability and contrary to common sense, either rises by 60 percent or falls by 40 percent. In his will the financier stipulates that his heirs are not to sell their portion of the team for one hundred years. The amount the heirs will receive depends, of course, on the number of years the team's value will have risen, but its mathematical mean is a whopping $1,378,000,000, while its mode, or the most likely value of (their portion of) the team, is only a paltry $13,000. Why?*

---

*The explanation for this disparity is that there is more "numerical room" for good luck to operate than there is for bad luck. The mean is an average over all possible hundred-year periods, and the astronomical returns associated with lucky hundred-year periods (in which there are many more years of 60 percent surges than of 40 percent declines) skew the returns upward. During those unlucky hundred-year periods in which the opposite occurs, however, the returns are bounded below by $0. In more detail: The team's value rises an average of 10 percent each year (the average of +60 percent and −40 percent). Thus after one hundred years, the mean of the investment is $100,000 × $(1.10)^{100}$, which is $1,378,000,000. On the other hand, the most likely result is that the team's value will rise in exactly fifty of the hundred years. Hence the mode (and the median) is $100,000 × $(1.6)^{50}$ × $(.6)^{50}$, which is only $13,000.

Bucks County and Environs
A Note on Maps and Graphic Games

The philosopher Ludwig Wittgenstein wrote that language consists of many different language games, brief exchanges, and situational colloquies that have their own idiosyncratic uses and rules that must become part of people's linguistic repertoire if they are to be said to know the language. Nowhere is this dictum better exemplified than in the daily newspaper, where the earnest narrative language of the news sections is juxtaposed with gossip, opinion, exhortation, analysis, lament, persuasion, and advice. Usually it is clear which language game is being played, but confusion is not uncommon. People who talk or write about a criminal trial "sending a message" to society, for example, are not really playing the legal game; sending messages is not the purpose of a trial.

A more pictorially oriented Wittgenstein might have extended his insights to graphics. Pie charts and histograms are the analogues of straight news reporting, but there are many other sorts of graphics that have distinct uses, operate under different rules, and must be interpreted accordingly.

Take a city or highway map obtained from any service station for a trivial example (in the literal sense of *trivial,* which is derived from the Latin word for "crossroads"). The fragment of one I'm looking at now happens to accompany a recent crime story in the paper; it may constitute evidence in the case. Let me ignore the case, however, and scrutinize the map, which indicates a number of less than stellar Points of Interest (POIs) near the crime scene. Are there many people who drive along a street or highway, note that one of these Points of Interest is only 8 miles off the road, and decide to visit it? Why, then, do such maps usually show these POIs and not other, more critical features? What about the crops, economic activity, and environmental characteristics of the surrounding area—wheat fields, factories, even landfills? Or bike and public transportation routes? Are county lines really significant to anyone besides county administrators? And do highways really become so sparse just on the other side of our state's border? With all their quirks, most of us learn how to use these maps, learn what to pay attention to and what to ignore.

The newspaper doesn't generally contain street or highway maps, but it does include tourist maps in the travel sections, and these require slightly different decoding skills. So do weather maps and organizational flowcharts, political cartoons and advertising layouts, lingerie ads and parade routes, business data and scientific illustrations, accident-scene sketches and crossword puzzles.* Maps of Bosnia or the Middle East require a whole course to decipher and sometimes seem to provide counterexamples to the theorem that any map can be colored with at most four colors. Incidence matrices, discussed in a later segment, provide an instance of a new graphics game. And photographs, of course, have their own logic, which is changing with the easy mutability that digitalization allows.

I'm always surprised at how much dexterity with a variety of language and graphics games is required to read the newspaper. It isn't a unitary interpretive skill, but a collection of specialized ones that ranges from decoding box scores to finding the strike price on your stock options. Isaiah Berlin quotes Archilochus: "The fox knows many things, but the hedgehog knows one big thing." The many short features in newspapers are for foxes. Like this segment, however, some of them are occasionally merely shaggy-dog stories with no particular point.

---

*Crossword puzzles may have lost some of their appeal because of the tempting ease with which reverse dictionary software and wild-card symbols can be used to solve them.

Ask About Your Mother-in-Law's Lladro

Explanations, Advice, and Physics

When did Myrtle punch her brother? Why did George and Martha divorce? Why is Oscar so withdrawn? What should be done before confronting him? Personal advice columnists answer these questions easily, but their answers often appear glib. Why didn't she have children? It was because she had been a mother to her own sickly and disturbed mother. Why doesn't she speak to me when we visit? Maybe she would if you first discussed her Lladro collection. Inane as some of these responses are, it's always impressed me that they're more illuminating than a biochemical or physical account of the same phenomena. Physics, chemistry, and biology may be more basic sciences than folk psychology, but Ann Landers is more likely than your local physicist to tell you how to deal effectively with your neighbors' meddling.

There is a distinction, thanks to Daniel Dennett, that is useful in this and other contexts and is most easily explained by reference to a computer. We take an *intentional stance* toward a computer when we impute desires, intentions, and goals to it and respond accordingly. This is what one does when playing chess against a machine. If I move my rook there, I reason that it, being rational in a sense, will use a knight fork against me and so I decide to do something else. An explanation of the computer's chess moves that invokes its aims and purposes is an intentional explanation.

By contrast, we assume a *design stance* or a *physical stance* toward a computer when we examine its software or hardware, respectively, in order to fashion an appropriate response. To do this with a chess-playing computer would be a futile, aeon-consuming task. Imagine trying to determine the machine's next move by looking at the reams of code in its program or by scrutinizing the myriad circuit connections among its chips. An explanation of the machine's moves that points to its program or electronics, respectively, is a design or physical explanation.

When applied to people, these distinctions help to clarify the contrast between actions that require intentional explanations (Myrtle punched her brother because he was teasing her) and those that demand only a physical explanation (Waldo fell down the elevator shaft). A physical account of the punching action (her closed hand traveled through this angle with that velocity and decelerated . . .) is clearly inappropriate. So is an intentional account of the fall (unless it was a suicide). Sometimes it's not clear which sort of explanation is called for. If George and Martha are huddled together, Martha may begin to cry. Is she responding to the emotional content of George's message, or did he blow onion powder in her eyes?

An obvious problem with intentional explanations is that, unlike more physical ones, they are as varied as the situations in which they arise. Many different human actions can correspond to the same physical movements, a revealing fact about the limited relevance of physics to human predictability. Suppose, for example, that you're searching for an explanation for why a man has just put his hand to the side of his head. In principle, the physical explanation is easy, unique, and uninformative: Neuron firings and muscle contractions brought about by a complex set of physiochemical phenomena caused the upper right appendage to move toward the lateral part of the uppermost central extremity.

The intentional explanation could involve any of many reasons. Maybe it was a windy day and the man was checking on the snugness of his toupee. Or he may be a political aide signaling his candidate to think before responding. Alternatively, he might be castigating himself for having overlooked something he's just now remembered. Or maybe he ate spicy food that is making his scalp itch. Possibly he's feigning nonchalance because a conversation he's engaged in is making him nervous. Or the gesture might mean nothing.

And touching one's hand to one's head is a very simple action. Collect twenty or so such actions with all their possible interpretations, put them before someone or some group with a distinct agenda, political or otherwise, allow for vagueness, omissions, distortions, and lack of context, and just about any interpretation could result. Some practitioners of literary hermeneutics make a living conjuring up outlandish accounts of this or that seemingly straightforward novel. Similar temptations exist for the advice columnist.

Of course, this distinction between intentional explanations and more physical ones says nothing about the appeal of advice columns, which derives in large part, I think, from the same source as does eavesdropping. This appeal is not unrelated to that of psychic and astrology columns with headlines like CAPRICORNS SHOULD BEWARE ASSOCIATES' GENEROSITY. The advice is open-ended and cryptic and occasionally provokes a degree of thinking and self-exploration that more definitive pronouncements do not. It is only the implicit claims to validity that these columns also make that are groundless. Still, I'd be opposed to the draconian suggestion that advice columns be prefaced with disclaimers that they are meant for entertainment purposes only, the way some astrology and psychic columns are. Such disclaimers might suggest that no other parts of the paper require them.

Garden Club Gala
Incidence Matrices on the Society Pages

W hether expressed or tacit, pronouncements like "Everybody was there" or "They're all doing it now" have always made me wince. Who exactly is everybody and how does it come to be known that they're all doing it? A passage from Boswell's *Life of Johnson* is apt:

> Sir, it is amazing how things are exaggerated. A gentleman was lately telling in a company where I was present, that in France as soon as a man of fashion marries, he takes an opera girl into keeping; and this he mentioned as a general custom. "Pray, Sir, (said I,) how many opera girls may there be?" He answered, "About fourscore." "Well then, Sir, (said I,) you see there can be no more than fourscore men of fashion who can do this."

Another aspect of articles in the society pages or in stories about political and entertainment figures that annoys me is the suggestion that "everybody" knows everybody else. I wish there were what mathematicians call an incidence matrix accompanying the story. An incidence matrix is a square array of numbers, each a 0 or a 1, indicating whether or not entities are connected (see diagram). If the story concerns a dozen people, for example, there would be twelve rows of twelve numbers, the number in the *i*th row and *j*th column being 1 if person i can initiate contact with person j and 0 if not. Status disparities ensure that a 1 in the *i*th row and the *j*th column does not always imply a 1 in the *j*th row and the *i*th column. (Don't call me; I'll call you.) Other relationships between pairs of people can be modeled in the same way.

Incidence Matrix

	#1	#2	#3	#4	#5	#6
Person #1	1	0	0	0	0	0
Person #2	1	1	0	0	1	0
Person #3	0	0	1	0	0	1
Person #4	0	1	0	1	0	1
Person #5	1	0	0	0	1	0
Person #6	0	0	0	1	1	1

A 1 in the *i*th row and *j*th column indicates
person #i knows where person #j is.

It's unlikely that such a matrix could be constructed by the writer of the society story, but if it could, various mathematical techniques could be applied to the matrix to yield further information. By multiplying this matrix by itself (in the special way in which matrices are multiplied), one could determine the number of two- and three-step communication paths from i to j and the identity of central figures in the group. One could also infer the existence of cliques and dominance relations in the group. Of course, some might prefer reading that "everyone was there."

Incidence matrices would be useful for other sorts of stories as well. Tools from graph theory and the theory of simplicial complexes might even provide unusual insights into the given relationships not otherwise obtainable. You might try constructing one for your friends and acquaintances.

A little exercise: Assume that a 1 in the *i*th column and the *j*th row in the diagram above indicates that i knows where j is hiding. You are person #3. How would you go about finding where person #1 is hiding; in other words, who would you find first who could help you find someone else, and so on?*

---

*The shortest way would be for you to find #6 first, which the diagram indicates you can do. Then #6 can help you find #5, who, in turn, can help you both find #1. Again, the incidence matrix indicates that this is possible.

Ten Reasons We Hate Our Bosses

Lists and Linearity

The Top 10 list has become a staple of newspapers, television, and magazines for a variety of reasons, the top ten being:

1. Ten is a common and familiar number, the base of our number system. Numbers are rounded to 10 or to multiples of ten or tenths. The resulting distortion, of course, need not have much to do with reality. We're told, for example, that we use 10 percent of our brain power, that 10 percent of us consume 90 percent of the world's resources, and that decades define us. (Is there anything more vapid than explanation by decade? In the free-love, antiwar sixties, hippies felt so-and-so; the greed of the eighties led yuppies to do such-and-such; sullen and unread Generation Xers never do anything.)

2. People like information to be encapsulated; they're impatient with long, discursive explanations. They want the bare facts, and they want them now.

3. The list is consistent with a linear approach to problems. Nothing is complex or convoluted; every factor can be ranked. If we do a, b, or c, then x, y, or z will happen. Proportionality reigns.

4. It's a kind of ritual. Numbers are often associated with rites, and this is a perfect example.

5. It has biblical resonance, the Ten Commandments being one of its first instances. Others are the ten plagues on the Egyptians, the ten days between Rosh Hashanah and Yom Kippur, the requirement that at least ten men be present for public prayer, and Joseph's ten brothers.

6. The list can be a complete story. It has a beginning: 1, 2, 3; a middle: 4, 5, 6, 7; and an end, 8, 9, 10. Many stories in the news are disconnected; the list is unitary.

7. It's easy to write; there is no need to come up with transitions. Or even complete sentences. The same holds for the 10, 50, AND 100 YEARS AGO TODAY fillers.

8. It's flexible and capable of handling any subject. Since there are never any clear criteria for what constitutes an entry on such a list, items on short lists can easily be split, and those on long lists can just as easily be combined.

9. Lists are widely read (or heard) and talked about, but don't require much room in the paper or much airtime.

10. People realize it's an artificial form and like to see if it's going to run out of good points before it gets to 10.

Stallone on Worst-Dressed List
Traits and Rates

A simple point about a common newspaper feature: The lists for best- and worst-dressed, for most admired and most hated, and for most Xish for any publicly appraised trait X can be determined by a certain mathematical product—a rough numerical measure of the trait ostensibly in question multiplied by the recognition rate of its possessor. (I doubt if the list makers conceive of it this way, however.) A celebrity can end up on the worst-dressed list, for example, merely because he or she wore an odd outfit once or twice and is known to tens of millions of people. The collection of possible candidates for these lists is effectively limited to a relative handful of celebrities.

The homeless person on the corner is certainly dressed worse than any celebrity, but he is known only to the people who step over him every morning. Likewise, anyone moderately controversial and known to virtually everyone might well appear on both the most admired and the most hated lists. More informative for national politicians and others whose recognition is near universal is merely to measure the percentages of people who view them positively and negatively.

These lists are just one of the many measuring protocols and unvalidated quizzes and surveys that misrepresent a characteristic. Another example is the perennial news story ranking the most livable cities in the country. The criteria used are as numerous as they are vague: cultural activities, parks and recreation, professional sports, crime, education, proximity to natural attractions, income, jobs, public transportation, traffic, air and water quality, and so on. These factors are then weighed, but by whom and by what standards? Someone with the artistic sensibility of Edward Hopper might give high marks to a Midwestern town with many deserted streets, dreary bars, and empty diners.

New Biography Fills Much-Needed Gap

Books and News

As an author of five books and a reviewer of dozens, I may be biased, but I think that books should be bigger *news* than they are. More than 50,000 are published annually, but only a small fraction of these are reviewed. Most of these 50,000+ books, even the most hackneyed, have received love and nurturance from their parents/ authors. Excepting those that fill much-needed gaps,* couldn't a larger percentage receive at least some sort of birth announcement/review?

Every baseball, basketball, and football game, whether at the professional, college, or high school level, is lavishly reported with statistics of every imaginable sort. Every gritty detail of murders, drug deals, and other abuses makes the paper. Every TV program on every cable channel has a brief synopsis in a weekly or monthly guide. Every minuscule variation in the stock price of hundreds of penny-ante companies is right there in the papers every day. I can't believe the readership for a daily stream of nationally syndicated, very brief reviews of new books would attract fewer readers than these features do. Besides, newspapers have a vested interest in a more literate reading public.

---

*I once heard an author characterize his own book this way during a television interview. Given the time constraints of most such interviews and the fact that the author is generally so intent on getting the gist of the book across, such blunders are almost inevitable, but rarely fatal. More worrisome to an author is the thought of a *reviewer* characterizing his or her book as filling a much-needed gap, say, between mathematics and journalism.

Those relatively few books that are reviewed sometimes receive a lavish amount of attention, much of it embarrassingly fulsome. There may be celebrity profiles not too dissimilar from the one sketched in an earlier section. Regarding blurbs, I note that if N positive ones are required to appear on a book's dust jacket, then, on average, there need be N × $^1/_P$ requests made to obtain them, where P is the percentage of prepublication reviewers who like the book. Thus, to find four positive blurbs for a book when only one out of six readers is likely to endorse it requires approximately twenty-four requests. Of course, many factors other than a book's quality affect this percentage.

Having written about lists in the two previous segments, I should say something about best-seller lists here. Unfortunately, there is little hard knowledge available. Not surprisingly, books on such lists become better known for being well known than for being widely read. Even if such a book is read by everyone who purchases it, for example, a best-seller that sells 100,000 copies is read by only .038 percent of Americans; some identifying blurb on the book may become familiar to hundreds of times as many people, however. Appearing on a list is certainly an important pane in the hall of media mirrors that is lined with newspaper interviews, magazine articles, personal appearances, and television talk shows. One bit of attention can feed so seamlessly into another that during the time that my book *Innumeracy* was on the best-seller list, I felt like a passenger on a very large surfboard riding over giant, almost sentient waves of publicity. Needless to say, luck, timing, and some quite nonlinear effects play a role in a book's making it to bestsellerdom.

In any case, more newspapers than ever are maintaining best-seller lists, and each has its own formulas and quirks that publishers try to exploit, if possible. The best-seller list put out by the *New York Times* is number one on the list of all best-seller lists for two reasons. It samples a larger number and a greater variety of outlets than do other lists, and it's published in the *New York Times*.

Which Way Mecca?

Religion in the Paper

In most newspapers religious *belief* is barely referred to. One must search long and hard for any mention of it, much less a serious news article. There are, however, many stories on the sociological relevance of religion: socially responsible clergy, libidinous priests, power struggles within denominations, charitable endeavors, fundraising drives, hate crimes, the political clout of the religious right or left, and so on.

A recent such story dealt with the Roper poll, which indicated in 1993 that almost one-fourth of Americans doubted the reality of the Holocaust. Now it turns out that this figure was a consequence of a question containing a very confusing double-negative construction. Rephrased, the question elicited the less distressing information that only 1 percent of Americans thought that it was "possible ... the Nazi extermination of the Jews never happened."

Another example of the way newspapers deal only glancingly with religion was an account in the *Philadelphia Inquirer* of the difficulty various mosques in the United States were having in deciding in which direction Mecca lay. The orientation of prayer, expensive buildings, and doctrine was at issue. Was one to face Mecca on a straight line through the earth or in the northeasterly direction of the shortest great circle route (as an airplane traveling there would)?

Recently a number of commentators, including Stephen Carter—whose book *The Culture of Disbelief* President Clinton was reported to have carried around for weeks—have bemoaned this superficial coverage of religion. They have argued that a subject as significant as religion is for many people should receive more and deeper reportage in the paper and elsewhere. An obvious counterargument is that centuries of simmering hostility among religious communities have probably left most of us justifiably chary of publicly discussing our beliefs or lack thereof; we have opted for a sometimes uneasy silence.

Were this to change and were substantive discussions of religious topics to take place in the paper and other media, people would have to allow for religious doctrines contrary to their own, as well as for antireligious beliefs, to be freely expressed. But when was the last time even a passing reference to the absence of any empirical evidence for God's existence (much less any particular religious tenet) appeared in the newspaper or on TV, and when were the words *superstition* and *alleged* attached to historico-religious events or the adjective *blasphemous* applied to rival religious claims? Even seemingly minor solecisms in religious stories can arouse great fervor. A *Time* magazine cover from a few years ago asked "Who Was Jesus?" and not, as some irate readers contended it should have, "Who Is Jesus?"

It would be wise, in my opinion, to continue our tacit embargo on public expressions of religious belief. I wouldn't want to see people on afternoon talk shows babbling about their faith. Even a devout agnostic such as myself would hate to see the simplistic trivialization that would likely ensue if religious shows and testimonials became common in the mainstream media. There is a fine line between public expressions of faith and aggressive declarations thereof, and religious tolerance is inversely proportional to the latter. Honoring my own counsel, I'll desist from further comment.

R. L. Vickler, 85, Aide to Truman
The Length of Obituaries

And then there are the obituaries, whose vertical slices into the past are a pleasing respite from the horizontal nowness of most newspaper articles. I remember skimming the obituaries even as a teenager, but now I always look to see if there are any names I recognize. If not, I glance at the headings and read the items that interest me. I generally check the ages and occupations to see if there is any connection between the two. Educators seem to live a long time, working-class people seldom receive announcements, and a disproportionate number of the deaths of middle-aged men in the arts are from AIDS.

I wonder about the relationships among the obituary's length, L; the deceased's achievements, A; his or her fame, F (which is largely independent of achievement); the interval between these and death, I; and the number of other "important" deaths that day, D. Maybe it's something roughly like $L = (A \times F^2)/\sqrt{(I \times D)}$. For the majority of the deceased, dying is the last newsworthy thing they can do. I'm not sure whether this is a depressing fact or not.

Finally, deaths possessing an element of incongruity stick with me. Adelle Davis, who sermonized on wholesome eating, died of cancer, a disease she attributed to having gorged on potato chips in her youth. The fitness guru and running advocate James Fixx succumbed to a heart attack. Perhaps my own trajectory will spiral into oblivion as a result of some egregious miscalculation of mine or a condition whose risk I have minimized. Would there be any lesson there? Of course not.

Conclusion

*The best lack all conviction, while the worst are full of
passionate intensity.*
  –WILLIAM BUTLER YEATS

In addition to the explicitly mathematical points I have made in the
preceding pages, there are two general sorts of conclusions I'd like
you to draw from this book. One concerns issues of journalistic
hygiene and what we can do to improve ours; the other, matters of
irreducible uncertainty and why we must learn to accept them. In
short: Always be smart; seldom be certain.

First, the positive prescription. I've argued that the set of standard questions journalists ask and readers want answered should be
enlarged. Besides Who, What, Where, When, Why, and How, it
should include How many? How likely? What fraction? How does
the quantity compare with other quantities? What is its rate of
growth, and how does that compare? What about the self-referential
aspects of the story? Is there an appropriate degree of complexity in
it? Are we looking at the right categories and relations? How much
of the story is independent of its reporting? Are we especially vulnerable to the availability error or to anchoring effects?

If statistics are presented, how were they obtained? How confi-
dent can we be of them? Were they derived from a random sample
or from a collection of anecdotes? Does the correlation suggest a
causal relationship, or is it merely a coincidence? And do we understand how the people and various pieces of an organization reported
upon are connected? What is known about the dynamics of the
whole system? Are they stable or do they seem sensitive to tiny per-
turbations? Are there other ways to tally any figures presented? Do
such figures measure what they purport to measure? Is the precision
recounted meaningful?

Without diminishing the civic and journalistic importance of these critical questions or the other mathematical issues raised herein, I end with a cautionary note that has also been sounded in a number of the book's segments. Even after reading with the greatest discernment, we will sometimes remain baffled by what we read for the irremediable reason that the world *is* baffling. Its sheer size, intricate connectivity, sensitive dependencies, self-referential tangles, random juxtapositions, and meaningless coincidences ensure this. Add the uncharted and nonlinear interactions within, between, and among disparate systems and you come up with complexity with a capital C. Not surprisingly, our responses to fads, fashions, politics, economics, and other events and processes beyond our understanding (or information-processing capacity) will be many and various, tendentious and controversial. Almost any bunkum has some partial validity, and we regularly read into the confusing mess what we want to see.

Clarity fading into murkiness suggests a sort of knowledge-based Peter Principle. Recall that this eponymous bit of folk wisdom asserts that people in a large organization tend to be promoted and advanced in their careers until they reach a position in which they perform poorly. At this level of relative incompetence, they usually linger for the balance of their careers. This may or may not be so, but it does provide us with a suggestive analogy. Substituting for the organizational hierarchy the fine structure of the natural and social sciences, commonsense psychology, and other schemes of explanation, we arrive at a somewhat similar observation.

Whether we admit it or not, it seems that we all tend to rise to our level of uncertainty. We master the easy links, the local correspondences, the ways to get by (of which there are many, or we wouldn't be able to function). New understanding develops, but we tend to keep pushing until we come up against social and physical phenomena that are too complex for us to grasp or foresee in any detail. At this point we hesitate, consult, banter, gamble, and ultimately proceed.

Immersed in this booming, buzzing, burgeoning network of information, we nevertheless often find that the answers we're most interested in remain beyond our complexity horizons. All crystal balls are cloudy or, in the case of newspapers, inky. Despite the promise (or menace) of mathematical prophecies of the future, we must frequently settle for the headlines of the present. That's why newspapers will always be new, why there will always be an element of romance in them.

Buy one and read all about it.

Bibliography

The bibliography for this book consists largely of the newspapers referred to in the text. The following books, among many others I could cite, elaborate on some of the points mentioned herein.

Casti, John L. *Searching for Certainty.* New York: Morrow, 1990.

Chaitin, Gregory J. *Algorithmic Information Theory.* New York: Cambridge University Press, 1990.

Cohen, Elliot D., ed. *Philosophical Issues in Journalism.* New York: Oxford University Press, 1992.

Cohn, Victor. *News and Numbers.* Ames: Iowa State University Press, 1989.

COMAP. *For All Practical Purposes.* New York: Freeman, 1994.

Davis, Philip, and Reuben Hersh. *The Mathematical Experience.* Boston: Houghton Mifflin, 1981.

DeLong, Howard. *Profile of Mathematical Logic.* Reading, Mass.: Addison-Wesley, 1971.

Dennett, Daniel C. *Brainstorms.* Acton, Mass.: Bradford, 1978.

————. *Consciousness Explained.* Boston: Little, Brown, 1991.

Dixit, Avinash K., and Barry J. Nalebuff. *Thinking Strategically.* New York: Norton, 1991.

Glieck, James. *Chaos.* New York: Viking Press, 1987.

Hofstadter, Douglas R. *Gödel, Escher, Bach.* New York: Basic Books, 1980.

————. *Metamagical Themas.* New York: Basic Books, 1985.

Kemeny, John G., J. Laurie Snell, and Gerald L. Thompson. *Introduction to Finite Mathematics.* Englewood Cliffs, N.J.: Prentice-Hall, 1980.

Kurtz, Howard. *Media Circus.* New York: Times Books, 1994.

Lem, Stanislaw. *One Human Minute.* New York: Harcourt Brace Jovanovich, 1986.

Lewis, H. W. *Technological Risk.* New York: Norton, 1990.

Manoff, Robert Karl, and Michael Schudson, eds. *Reading the News*. New York: Pantheon, 1986.

Moore, David, and George McCabe. *Introduction to the Practice of Statistics*. New York: Freeman, 1993.

Paulos, John Allen. *Innumeracy*. New York: Vintage, 1990.

————. *Beyond Numeracy*. New York: Vintage, 1991.

Ross, Sheldon. *First Course in Probability*. New York: Macmillan, 1994.

Ruelle, David. *Chance and Chaos*. Princeton, N.J.: Princeton University Press, 1991.

Stewart, Ian. *Does God Play Dice?* Cambridge, Mass.: Blackwell, 1989.

Sutherland, Stuart. *Irrationality: The Enemy Within*. London: Constable, 1992.

Tversky, Amos, and Daniel Kahneman. *Judgment Under Uncertainty: Heuristics and Biases*. New York: Cambridge University Press, 1982.

Index